Fly Patterns

Fly Patterns

An international guide

Taff Price

Illustrations by George Thompson

WARD LOCK LIMITED · LONDON

To John Veniard,
whose help, encouragement and friendship over many years
has made this book possible.

First published in Great Britain in 1986
by Ward Lock Limited, 8 Clifford Street
London W1X 1RB, an Egmont Company

Printed in Spain by Graficromo

Text filmset in Linotron Palatino
by Dorchester Typesetting Group Ltd.

British Library Cataloguing in Publication Data

Price, Taff
Fly patterns: An international guide
1. Flies, Artificial
I. Title
799.1'2 SH451

ISBN 0-7063-6362-0

Acknowledgments

Luis Anntunez, Madrid, Spain
Jack Blackman, Natal, South Africa
Norbert Eipfltauer, Vienna, Austria
Marjan Fratnik, Milan, Italy
Roland Heriegel, Interfly, Zurich, Switzerland
Bob Lewis, Zomba Fishing Flies, Malawi
Luciano Maragani, Milan, Italy
Darrel Martin, Tacoma, Washington, USA
Roman Moser, Traun River Products, Siegsdorf, Austria
Moc Morgan, Dyfed, Wales
O. Mustad & Son Ltd, Peterlee, Co Durham, UK
The Orvis Company, Hampshire, UK
Partridge of Redditch, UK
John Smith, Brookwood, Surrey, UK
Stef Stefferson, Tranbjerg, Denmark
Barry Unwin, Fulling Mill Flies, Kenya
Peter Veniard, E. Veniard Ltd, Thornton Heath, Surrey, UK
Dr Bozidar Voljc, Slovenia, Yugoslavia
John Wilshaw, *Trout & Salmon* magazine

George Thompson for the excellent illustrations
Helen Douglas-Cooper of Ward Lock for the editing

Contents

Introduction

The world of the artificial fly is a colourful and absorbing aspect of the gentle art of fly-fishing. The creation, with simple materials such as a pinch of fur and a feather or two, of an artifact that can outwit nature by fooling a trout, holds a fascination for many fishermen. Fishing is no longer just a matter of catching fish. It becomes something more; a clever deception, the fooling of a wild creature into believing that the concoction at the end of the leader is a natural insect, a potential meal, and worthy of the fish's attention.

The capture of fish for food must go back at least 10,000 years. Fishing as a sport, rather than as a basic requirement of life, was practised by the ancient Egyptians, as their tomb paintings illustrate. However, no one knows exactly when man began to produce these feathery creations in order to fool a fish.

The first definite reference to the artificial fly occurs in the work of the Roman writer, Aelian, around AD 250. Before then, references to what could have been artificial flies were made by Homer, in about 1000 BC. And Theocritus, in 300 BC, referred to 'The bait fallacious', the very word 'fallacious' having that ring of deception about it. However, it is Aelian who first described an artificial fly. He wrote of the Macedonians in the area close to the river Astracus, and how on that river the natives caught fish by means of an artificial fly. 'They have planned a snare for the fish and get the better of them by their fisherman's craft. They fasten red wool around the hook, and fit onto the wool, feathers that grow beneath a cock's wattles and which in colour are like wax.'

If these early fly-fishers wound the cock's feathers palmerwise around the hook, and it is possible that they did, then we have a direct link to the present day with the popular stillwater fly, the Soldier Palmer. It is not inconceivable that this primitive form of fly-fishing had been going on centuries before it was recorded by Claudius Aelian.

In England, in the year 1496 at the 'Syane of the Sonne in Flete Strete', a printer, Wynkyn de Worde, a former employee of William Caxton, printed a small pamphlet, *The Treatyse of Fysshynge with an Angle*. Ten years earlier at the Priory of Sopwill near St Albans, the Prioress, the good Dame Juliana Berners (or Barnes as it is sometimes spelt), had published a book, *The Bokys of Hauking and Huntyng and of Coot Armuris*.

The book was one of the earliest examples of English printing. This work and the subsequent *Treatyse* were combined in one volume some years later. Eminent scholars feel that the *Treatyse* was in all probability written much earlier, around 1450, and there are some who believe that most of it was taken from earlier manuscripts that could well have been French in origin. It is unlikely that the good Dame had anything to do with the actual *Treatyse* at all, but over the centuries Britain's first work on fly-fishing has become associated with her as there is no positive proof it was anyone else.

The *Treatyse* provided a list of twelve artificial flies, and the time to fish them. And despite some references to earlier texts on fly-fishing, it was this list that influenced fishermen's choice of flies for many years to come. And other writers in later years took freely from it.

The next book of importance was probably *The Book of Astorga* (1624) attributed to Juan De Bergara. It pre-dates many of the famous English fishing books, such as those by Barker, Walton and Cotton. *The Book of Astorga* listed the flies monthly for the area of Leon in Spain.

From its early beginnings in Macedonia fly-fishing and fly-tying have spread to all corners of the world. And in countries that did not have indigenous trout species, either brown or rainbow trout were introduced into their lakes and rivers. Ex-patriate tea-planters, district officers and military men carefully imported ova and fry and introduced them into the rivers and lakes of South Africa, Kenya, Malawi, New Zealand, Australia, Northern India, Ceylon, and many more places. Other afficionados of the fly rod introduced fighting rainbows and browns into the waters of Chile and Argentina.

From the cold north in Iceland right down to the Antipodes, man goes out to seek the spotted fish with a fly rod, and in all these countries there are those who sit at a fly vice, creating miniature works of art to fool the wary fish.

Flies from the Treatyse of Fysshynge with an Angle

March

The Dun Fly

Body of dun wool; wings of partridge. This fly could well be the March Brown.

Another Dun Fly

Body of black wool; wings of the blackest drake; jay under wings and tail. Some believe that this fly was the Large Dark Olive, but with its overall black appearance it is more likely to be an early Black Gnat of the Bibio species.

April

The Stonefly

Body of black wool with yellow under the wings and tail; wings of a drake. As its name implies this is a stonefly imitation.

The Roddyd Fly

A reddish wool body, with a black silk rib; wings of a drake and a red capon's hackle. This fly is thought to be the Red Spinner.

May

The Yellow Fly

Body of yellow wool; wings of red cock hackle and of the drake. This fly could well be the mayfly *Ephemera danica*, or the Yellow May Dun; or even the stonefly Yellow Sally.

The Black Louper

Body of black wool; peacock herl from an eye feather and wings of a red capon with a blue head. A looper caterpillar of some species, this could well be the original palmer fly.

June

The Dun Cut

Body of black wool with yellow either side; wings of buzzard bound with hemp. Almost definitely a sedge/caddis of some sort.

The Maure Fly

Body of dark wool and the wings of the darkest wild drake. This is likely to have been the Alder Fly.

The Tandy Fly

Tan wool body; wings from the speckled feathers of the wild drake. Many believe this to be a Cowdung Fly.

July

The Wasp Fly

Body of black wool ribbed with yellow silk; wings of buzzard. An imitation of a wasp, or perhaps a hover fly.

Shell Fly

Body of green wool with peacock herl; wings of a buzzard. Possibly another sedge fly.

August

The Drake Fly

Body of black wool ribbed with black silk; wings of speckled feather of the black drake with the black head. It cannot be said with any certainty what this fly represented.

Trout flies

DRY FLY PATTERNS

Dry flies, as the name indicates, are flies designed to be fished on the surface of the water. They are created to imitate a wide number of different insects.

Newly-hatched mayflies, floating on the surface to dry their wings before their first flight, are copied by a number of different artificial flies. This stage in the insect's life is known as the dun in angling parlance; more scientifically, as the sub-imago.

The female mayflies returning to the water to shed their eggs are known as spinners; after ovipositing, lying with wings outstretched in the surface film, they are called spent spinners or spent gnats. There are imitations to emulate these two stages in the mayfly's life.

Caddis, or sedge flies as they are often called, with their roof-like wings, and stoneflies, one of the most primitive of aquatic insect forms, are all copied by the fly-dresser, and receive the dry-fly treatment.

Apart from the various aquatic insects, there is a very large number of insects that find themselves in or on the water by accident rather than design. They can be blown there by the wind, they can fall off overhanging vegetation, or they may just lumber out of their natural element. All these insects are embraced by the term terrestrials, and form an important element in the diet of the trout. Beetles, grasshoppers, flies such as the black gnat and even the industrious ant, are such terrestrials, and receive attention at the fly-tying vice.

Sometimes dry flies are created with no particular insect in mind. Such artificials are termed as 'fancy' or attractor flies; the Wickham's Fancy is such a fancy fly.

Who was the first man to cast a dry fly? The answer is lost in the dim and distant mists of angling time. The earliest-recorded fly-fishermen were probably dry-fly fishermen. With their long, willowy poles and lines made of horsehair or some other natural fibre, which would have been tied to the top of the poles because there was no such thing as a fishing reel in those days, and their crudely tied flies, they dapped the surface in order to tempt the spotted fish of the River Astracus to rise from the depths.

Throughout early angling literature, references were made to the correct presentation of the fly. In all cases, these referred to the wet fly. Many of these writers, however, observed that as their wet flies landed delicately on the water's surface, they were often taken by the trout before they had time to sink.

There is a subtle difference between fishing a wet fly dry, and tying up a feathered lure specifically for fishing on the surface. It was James Ogden of Cheltenham, UK, who claimed to have originated the first artificial dry fly, around the year 1839. Another two firsts chalked up against Ogden's name were the first detached-body dry fly; and the modern method of dry-fly winging using the 'V' configuration, formed by slips of feather taken from a right and left wing-feather quill.

Having given credit to James Ogden for the dry fly's creation, although it cannot be definitely proved either way, we next come to a man whose name is a byword in dry-fly fishing, Fredric Halford. In the year 1886, with the help of his good friend G. S. Marryat, another eminent Victorian fly-dresser and fisherman, Halford published his book *Floating Flies and How to Dress Them.* So thorough was this book in every aspect, that its dictates are followed to this day.

On many rivers, there is an all-important, almost sacrosanct rule: 'upstream dry fly' only. In the past, adherents to the discipline of the dry fly applied it to all trout fishing. However, most anglers, on stillwaters in particular, will fish the dry fly when required and will change to a wet fly or nymph as the conditions or the quarry dictate.

Many dry flies are tied on up-eyed hooks, which are manufactured solely for this purpose. However, it makes little difference whether the eye is up or down, provided that the hook is fine enough. Nevertheless, some countries never resort to the up-eye for their dry-fly patterns. The illustrations here show dry flies tied on up- and down-eyed hooks.

The patterns given in this section come from all corners of the fly-fishing world. Some have been fished for over a hundred years and are tied with traditional materials; others are more modern, utilizing man-made fibres and modern ingenuity.

Greenwell's Glory

This is a traditional fly that has stood the test of time, and is regarded as a classic. Devised by Canon William Greenwell and tied by James Wright, this fly is still universally used by anglers all over the fly-fishing world. It can represent many of the different species called olives.

Hook 12-16.
Thread Primrose.
Tail None (however, sometimes given a tail to aid floatability).
Body Primrose tying silk well waxed to produce a shade of olive.
Rib Gold wire.
Hackle Greenwell (light furnace).
Wing Blackbird substitute.

Black Gnat

There are a number of natural flies covered by the term 'black gnat'. These insects are generally terrestrial flies of the Bibio species, though some anglers consider the water-loving Empid flies to be the true black gnat. Whichever is the case, it does not alter the fact that the Black Gnat is found in the fly-boxes of most anglers.

Hook 12-16.
Thread Black.
Tail None (but, like the Greenwell, a tail is sometimes added).
Body Black silk.
Rib Silver wire.
Hackle Black cock.
Wing Starling or grey duck.

Wickham's Fancy

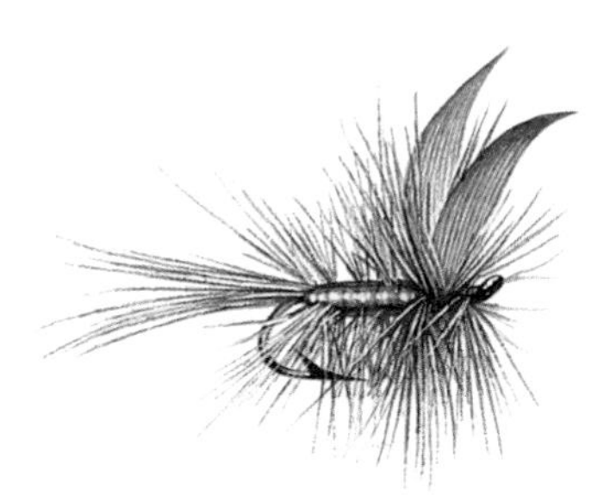

Another fly from gentler times that still retains a degree of popularity. The Wickham's does not represent any specific natural fly, but is a fancy dry fly. Though used extensively on rivers, a hackled version is used with some success by many British stillwater anglers.

Hook 12-16.
Thread Brown.
Tail Ginger or light red cock hackle fibres.
Body Flat gold tinsel.
Rib Oval gold tinsel, wire in smaller sizes.
Hackle Ginger or light red, tied palmer-style.
Wing Grey duck.

Blue-winged Olive

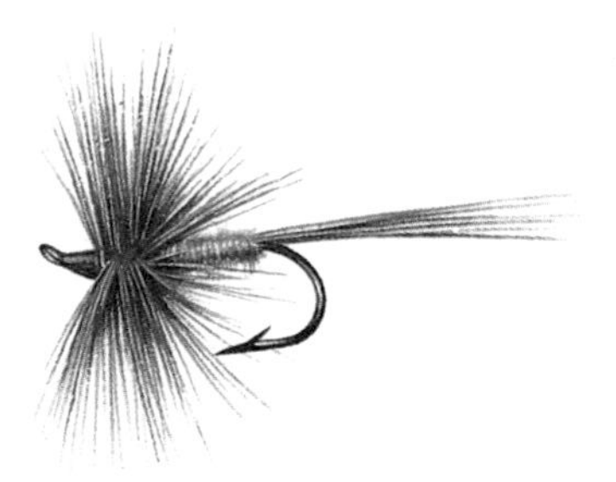

The blue-winged olive (*Ephemerella ignita*) is a widely distributed fly. It is found on wild streams as well as on the more sedate chalk streams. It is easily recognized as it possesses three tails; other olives have only two. This is an important pattern. The dressing given here is American in origin.

Hook 4-16.
Thread Olive.
Tail Dark dun hackle fibres.
Body Grey olive fur.
Rib None.
Hackle Dark dun.
Wing Dark blue dun hackle tips set upright.

Olive Dun

The term 'dun' is the fisherman's name for the first winged stage in an ephemerid's life. The more correct term is sub-imago. The artificial Olive Dun is a pattern that can imitate many of the natural olives. There are a number of different dressings for this fly.

Hook 12-14.
Thread Olive.
Tail Medium olive cock hackle fibres.
Body Cock hackle quill.
Rib None.
Hackle Medium olive cock.
Wing Grey starling.

Dark Olive

The large dark olive (*Baetis rhodani*) is one of the commonest members of the Ephemeroptera (mayflies) of the British Isles. It is found in all parts of the UK, and closely allied species are found throughout the USA and Europe. Good hatches can be expected early in the season, with further hatches later in the year.

Hook 12-14.
Thread Olive.
Tail Dark olive cock hackle fibres.
Body Dyed olive goose herl.
Rib Fine gold wire.
Hackle Olive cock.
Wing Starling or grey duck.

Rough Olive

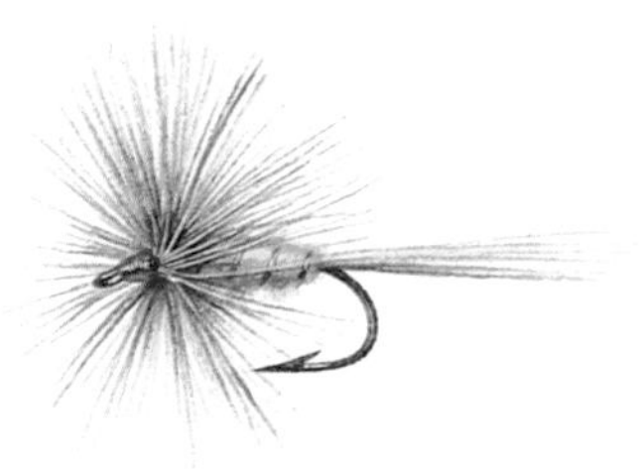

Another imitation of the Dark Olive. There is little to choose between this fly and the Dark Olive; both are considered to be excellent fish-takers by their respective adherents.

Hook 12-14.
Thread Olive.
Tail Brown olive cock hackle fibres.
Body Brown olive seal's fur.
Rib Gold wire.
Hackle Brown olive cock.
Wing None.

Alder

The natural alder (*Sialis lutaria*) is on the wing in the UK during the months of May and June. Many eminent writers of the nineteenth century wrote in fine praise of this artificial fly, but I must admit I have had very little success with it. However, I do know of others that have taken better than normal trout on a dry Alder.

Hook 12.
Thread Black.
Tail None.
Body Magenta-dyed peacock herl.
Rib None.
Hackle Black.
Wing Mottled brown hen.

Black Ant

This is a once-a-season fly; or if you are lucky, twice. The flying ant falls onto the water in large numbers in late summer, exciting the trout to selective feeding. This particular pattern was devised by Barry Kent, an expert English fly-dresser who evolved this fly while living in South Africa. It has worked well for me on both river and stillwater, taking both trout and grayling.

A red version can be tied using reddish brown suede chenille, and a yellow version using cinnamon yellow suede chenille.

Hook 14-16.
Thread Black.
Tail None.
Body Black suede chenille.
Rib None.
Hackle A few fibres of cock pheasant tail, tied in the middle.
Wing None.

Lunn's Particular

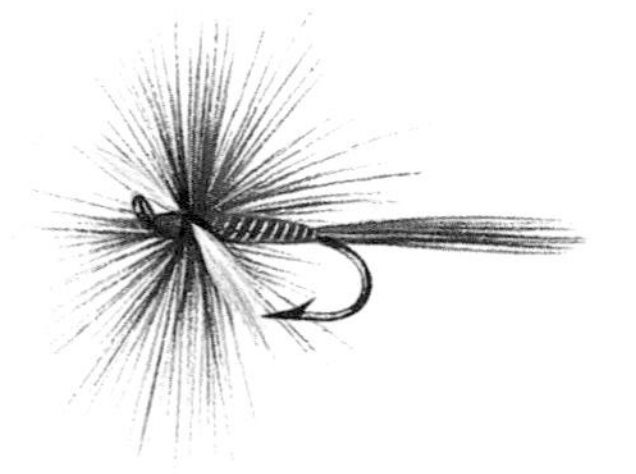

Of all the rivers in the world fabled for their fish and fishing, it is perhaps the meandering River Test of Hampshire that holds the crown of King of the Waters. It is the Mecca of most fly fishermen, few of whom will ever tread the banks of such a hallowed stream. W. J. Lunn was the keeper of the Houghton Club stretch for forty-five years, caring for his river from 1887. This fly is his creation; it first saw the light of day around 1916 and it has continued to catch fish since that time. It is a classic fly in all senses of the word. The care of the stream is still in the hands of the Lunn family, creating an unbroken tradition. This fly is a good imitation of the Olive Spinner.

Hook 14.
Thread Brown.
Tail Natural red cock hackle fibres.
Body Undyed stalk of a Rhode Island Red cock hackle.
Rib None.
Hackle Natural red cock hackle.
Wing Two medium dun cock hackle points, tied spent.

Coch-y-Bonddu

The June bug, field chafer and coch-y-bonddu are all names given to a little rotund beetle (*Phylopertha horticola*). As one of its common names implies, it is found in large numbers during the month of June. It is in greatest evidence in the Celtic fringes of the British Isles. Other imitations for this beetle are Marlow Buzz, Eric's Beetle, and Little Chap.

It would be a foolish angler indeed who did not have a few patterns of this fly if he was contemplating fishing the wilder parts of the country during the month of June.

Hook 12-14.
Thread Black.
Tail None.
Body Tip of flat gold tinsel, bronze peacock herl.
Rib None.
Hackle Dark furnace.
Wing None.

Cranefly

A large number of the family Tipulidae are terrestrial insects. However, a fair proportion are aquatic; that is to say, their larval and pupal stages are spent in water. Both types are taken readily by the trout.

Late-season fishing on the large reservoirs with the floating daddy longlegs is perhaps one of the highlights of the angling calender. There are many artificials tied to imitate this spindle-shanked fly, but this particular dressing has proved to be constantly effective.

Hook Standard size 10 or long shank 12.
Thread Black or brown.
Tail None.
Body Cock pheasant tail fibres.
Legs Cock pheasant tail fibres, knotted.
Rib Fluorescent green floss.
Hackle Natural red or cree.
Wing Cree hackle points.

Grey Duster

This fly can best be described as a broad-spectrum dry pattern, imitating nothing specific and yet imitating whatever the trout thinks it is. There are some who maintain that it is a fair representation of a moth of sorts. Others believe that it is effective during a rise to chironomid midges. The Grey Duster is a killing fly for both running and still waters.

Hook 12-16.
Thread Black or brown.
Tail None.
Body Rabbit fur mixed with the blue underfur.
Rib None.
Hackle Well marked badger.
Wing None.

Sherry Spinner

A fly tied to imitate the egg-laying stage of the blue-winged olive. Some prefer this pattern to the Orange Quill (overleaf). A well-known brand of sherry owes its name to this fly; or perhaps it is the other way around.

Hook 14.
Thread Brown (light).
Tail Honey dun cock hackle fibres.
Body Sherry-coloured floss, or a mixture of seal's fur to give sherry colour.
Rib Fine gold wire.
Hackle Honey dun.
Wing Optional: two blue dun hackle points tied spent.

Orange Quill

This fly is the creation of the famed father of English nymph fishing, G. E. M. Skues, who fished well into his eighties on the chalkstreams of Hampshire. He questioned and antagonized many of the dry-fly-only dogmatists of his time. Yet although he was a controversial figure in his day, he was respected by all, even though he must have appeared a heretic to many. He found this particular dry fly to be extremely effective when the blue-winged olive was on the water.

Hook 14.
Thread Hot orange.
Tail Natural red cock hackle fibres.
Body Orange quill; the original called for condor, ostrich is acceptable.
Rib None.
Hackle Natural dark red cock.
Wing Pale starling.

Ginger Quill

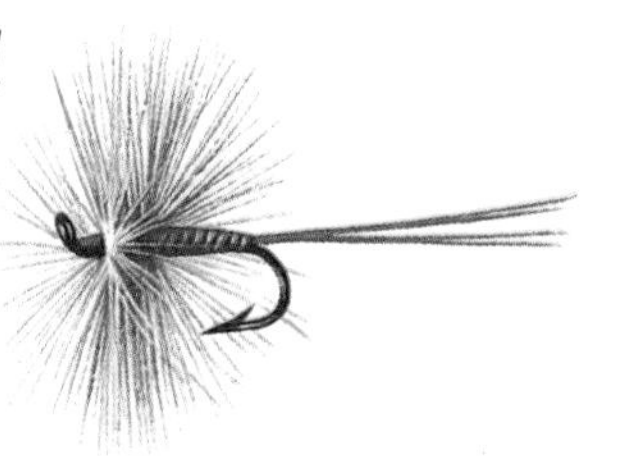

A favourite fly of many anglers, used on many of the West Country (UK) rivers, this is in essence a lighter version of the Red Quill. It is thought to imitate such insects as the pale wateries and other medium olives.

Hook 12-16.
Thread Black or brown.
Tail Ginger cock hackle fibres.
Body Stripped peacock quill from the eye of the feather.
Rib None.
Hackle Ginger cock hackle.
Wing Grey starling.

Red Quill

A sister fly of the Ginger Quill, this pattern can be used at any time. It is effective during a rise to iron blue duns and other olives. It does not imitate any specific insect and therefore falls into the broad-spectrum category of dry flies. Some anglers prefer to fish a wet version.

Hook 14-16.
Thread Black.
Tail Natural red cock hackle fibres.
Body Stripped peacock quill from eye feather.
Rib None.
Hackle Natural red cock.
Wing Starling.

Gold-ribbed Hare's Ear (GRHE)

This is one of my favourite flies; it would appear in my fly-box for any corner of the world. Though the pattern depicted is for a dry fly, it can be fished wet or as a nymph. It can be very effective fished in the surface film as a still-born olive. On British reservoirs, long-shanked, weighted versions have proved very successful for both rainbow and brown trout. An American nymph pattern appears later in the book.

Hook 12-14.
Thread Black or brown.
Tail Guard hair fibres from a hare's face.
Body Hare's ear.
Rib Gold wire.
Hackle Fibres picked out to form the hackle.
Wing None (sometimes a grey starling wing is added).

Blue Dun

This is a fly with a long lineage going way back to the time of Charles Cotton. Each part of the British Isles has its Blue Dun pattern, some winged, some hackled, some wet, some only fished dry. It is not really certain what natural insect the Blue Dun imitates, however, Alfred Ronalds said it changes into the Red Spinner so we can assume that it is yet another fly that represents the large dark olive.

Hook 14.
Thread Primrose.
Tail Blue dun cock hackle fibres.
Body Mole fur.
Rib None.
Hackle Blue dun cock.
Wing Grey starling.

Hardy's Favourite

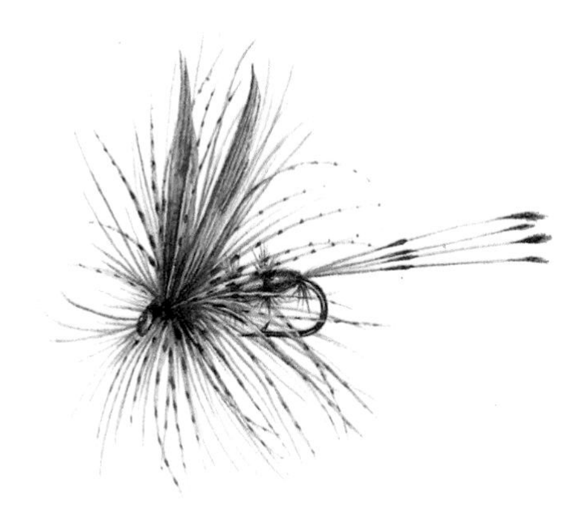

Like flies such as the Wickham's Fancy and the John Storey, this pattern is a fancy dry fly representing nothing in particular, but the trout find something edible about its appearance. It has also been used as a seatrout fly.

Hook 12-14.
Thread Black.
Tail Fibres of brown mallard feather.
Body Peacock herl.
Rib Red silk.
Hackle Dark partridge.
Wing Brown turkey.

French Partridge Mayfly

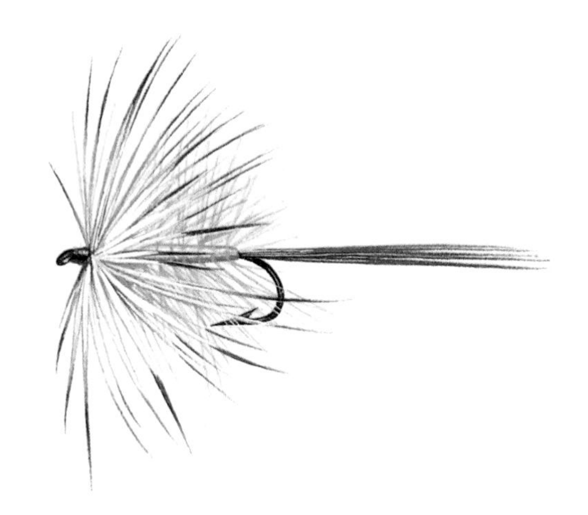

There are a number of different dressings of this fly. All of them use the barred hackle from a French partridge. This fly is one of my favourites.

Hook Long shank 12.
Thread Black or brown.
Tail Cock pheasant tail fibres.
Body Natural raffia.
Rib Gold wire.
Hackle Palmered olive cock down the body, French partridge feather at the head.
Wing None.

Walker's Mayflies

The late Richard Walker, whose contribution to modern fly fishing is unequalled, created these three mayfly patterns. The style of fly is known as 'Straddlebug' and has a definite Irish mayfly influence in the style of dressing. Note the colour orange in one of them. This colour holds a great attraction for both river and lake trout.

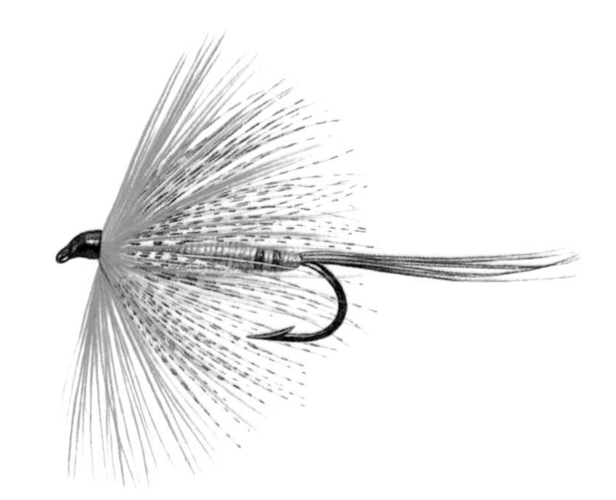

Hook Long shank 8-10.
Thread Brown.
Tail Cock pheasant tail fibres.
Body Buff turkey fibres.
Rib Two turns of cock pheasant tail at rear of hook.
Hackle Green cock, speckled duck.
Wing None.

Hook Long shank 8-10.
Thread Brown.
Tail Cock pheasant tail fibres.
Body Buff turkey fibres.
Rib Two turns of cock pheasant tail at rear of hook.
Hackle Hot orange, green, French partridge.

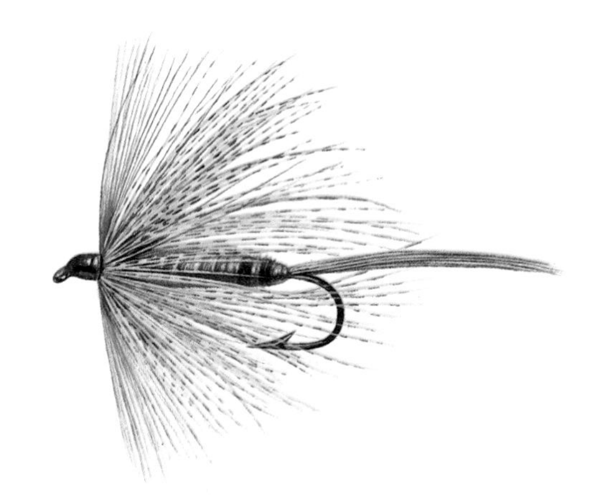

Hook Long shank 8-10.
Thread Brown.
Tail Cock pheasant tail fibres.
Body Buff turkey fibres.
Rib Two turns of cock pheasant tail at rear of hook.
Hackle Speckled duck dyed green, ginger.
Wing None.

Grey Drake

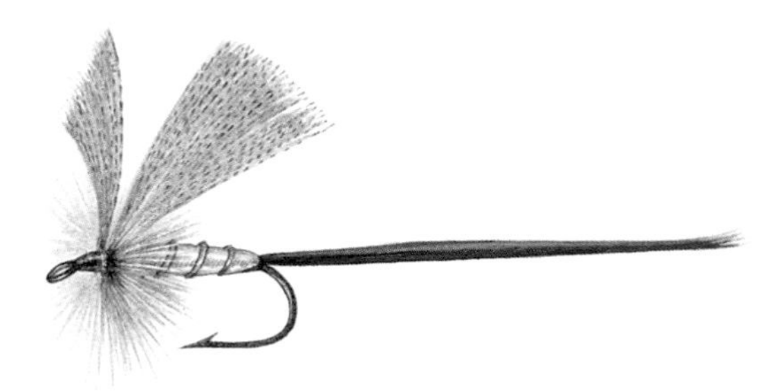

The term 'drake' originates from the feathers used in the wing of this fly. It is tied in the style termed 'fan wing', which is self-explanatory. Though looking realistic to the angler's eye, there is a hardening of opinion in the UK against this style of mayfly. Though I prefer the hackled type of fly, there are still many who swear by this style of dressing.

Hook Long shank 12.
Thread Black or brown.
Tail Fibres from a cock pheasant tail.
Body Natural raffia, or white silk.
Rib Black silk.
Hackle Badger.
Wing Grey duck body feathers, tied fawning.

Yellow Drake

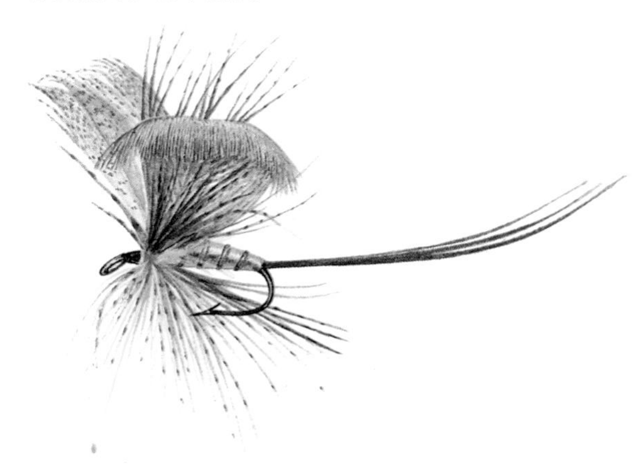

Yet another colour for the mayfly. Why yellow? Your guess is as good as mine. When floating amidst a host of naturals, this imitation stands out very obviously. Perhaps the trout take it for that reason.

Hook Long shank 12.
Thread Brown.
Tail Cock pheasant tail fibres.
Body Pale yellow silk.
Rib Black silk.
Hackle Yellowish olive.
Wing Duck breast feathers dyed yellow.

Green Drake

An olive version of the Grey Drake. It is thought that the Grey Drake represents the female dun of *Ephemera danica*, and the Green Drake the male of that species.

Hook Long shank 12.
Thread Olive.
Tail Cock pheasant tail fibres.
Body Olive floss.
Rib Gold wire.
Hackle Olive.
Wing Duck breast feathers dyed olive, tied fanwing.

Shadow Mayfly

This fly is the creation of Peter Deane, one of Britain's leading fly-dressers. It is a very impressionistic pattern which kills a lot of fish on the Hampshire chalkstreams. It has also been used as a dapping fly.

Hook Long shank 10-12.
Thread Black.
Tail None.
Body None as such; the hackle provides the body.
Rib None.
Hackle Palmered grizzle.
Wing Ginger hackle tips clipped at the top.

Hawthorn

Sometimes spelt Hawthorne, this fly is based on the natural, *Bibio marci. Marci* comes from the fact that it was supposed to hatch out on or around St Mark's day. This terrestrial fly is an essential pattern for the month of May in most parts of the UK. When the hawthorn blossom is out, you usually find the hawthorn fly on the wing.

Hook 10-14.
Thread Black.
Tail None.
Body Black-dyed goose fibres.
Rib None (some patterns use fine silver wire).
Hackle Black cock.
Wing Grey duck.
Legs Two feather strands left trailing.

Nevamis Mayfly

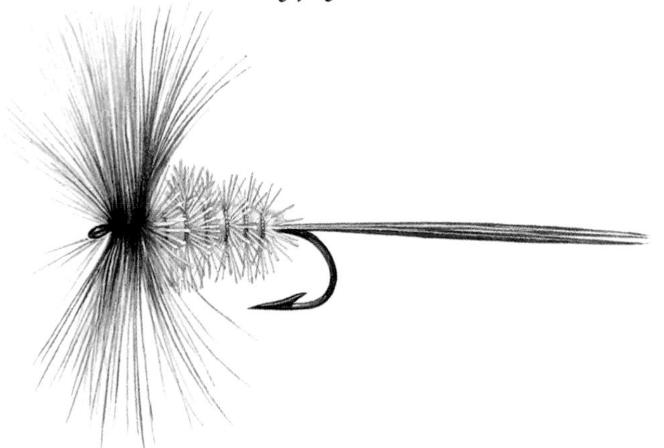

John Goddard's imitation of the mayfly, with good floating properties used by Goddard to great effect during the annual hatch of mayflies on the River Kennet.

Hook Long shank fine wire 8.
Thread Yellow.
Tail Cock pheasant tail fibres.
Body Cream seal's fur.
Rib Oval gold.
Hackle Palmered honey cock wound down the hook and clipped ¾ in (19mm) at the head down to ⅛ in (3mm) at the tail. Small furnace hackle at the head.
Wing Pale blue dun cock hackle fibres set in a 'V' upright.

Cinnamon Sedge

One of the most popular of the sedge or caddis fly patterns. This artificial can imitate a number of light brown sedges apart from the actual cinnamon sedge (*Limnephilus lunatus*). It emerges in June and can be found in quite large numbers, flying at dusk or late afternoon.

Hook 12-10.
Thread Brown.
Tail None.
Body Cinnamon turkey herl.
Rib Gold wire (optional).
Hackle Light ginger.
Wing Cinnamon hen or turkey wing quill.

Walker's Sedge

This is a large fly tied to represent the large red sedge and other large caddis flies. The addition of a small butt of fluorescent wool or silk at the tail end of the fly enhances its killing properties.

Hook 10-8.
Thread Black.
Tail None.
Body First a small tag of fluorescent arc chrome wool; ostrich herl (chestnut colour, clipped short).
Rib None.
Hackle Two natural red cock.
Wing A bunch of natural red cock hackle fibres.

Orange and Gold

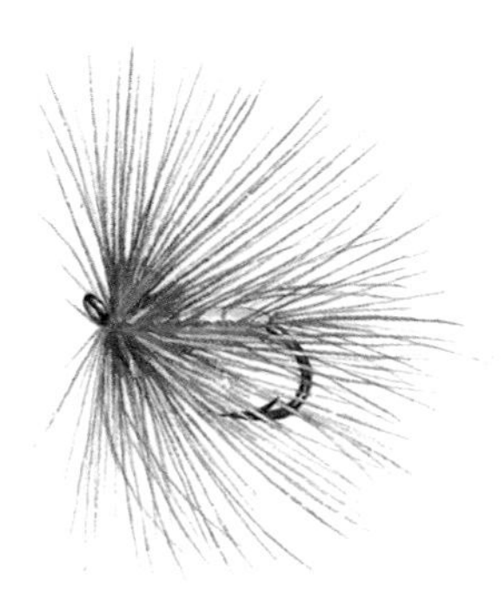

I was first shown this fly by one of the anglers who fish those large Midland reservoirs, Grafham and Rutland. It is particularly effective when trout are feeding on floating snails. The colour orange perhaps emulates the reddish/orange colour that shows through the infant snail's shell when viewed against the light. This colour is caused by the haemoglobin present in many of the pulmonate snails.

Hook 12-14.
Thread Brown or orange.
Tail None.
Body Fait gold lurex.
Rib None.
Hackle Palmered orange-dyed cock hackles.
Wing None.

Welsh Partridge

This can be used as a wet or dry fly and can well represent the natural claret dun (*Leptophlebia vespertina*). It can also work well as a March brown imitation. The fly was created by Courtney Williams, author of *A Dictionary of Trout Flies and of Flies for Seatrout and Grayling*.

Hook 12-16.
Thread Black.
Tail Two strands from a partridge tail.
Body Claret seal's fur.
Rib Fine gold oval.
Hackle Stiff claret cock with brown partridge in front.
Wing None.

Bumble Bee

There are times when, no matter what fly we use, we cannot seem to tempt the trout to take. Quite often in these situations, large, incongruous flies like the Bumble Bee defy all reason and cause the fish to take. Many times I have seen trout rise and take natural bumble bees when they have fallen on the water. This fly was devised by the late Richard Walker.

Hook 6.
Thread Black.
Tail None.
Body White, black and amber ostrich herl.
Thorax Black ostrich herl.
Legs Black cock pheasant tail fibres knotted.
Rib None.
Hackle None.
Wing Grizzle hackle points.

Blue Upright

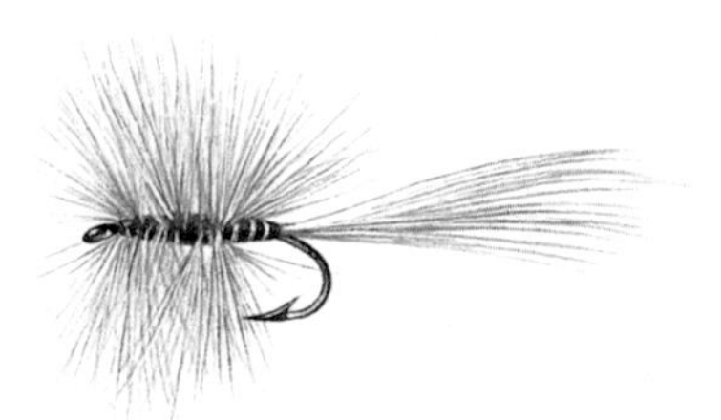

A favourite fly from the county of Devon. It was created by R. S. Austin, famous for the Tups Indispensable. This fly is a good floater for the rough streams as the hackle extends from the shoulder to the eye.

Hook 10-14.
Thread Purple.
Tail Steely blue cock hackle fibres.
Body Undyed peacock herl, stripped.
Rib None.
Hackle Steely blue cock hackle.
Wing None.

John Storey

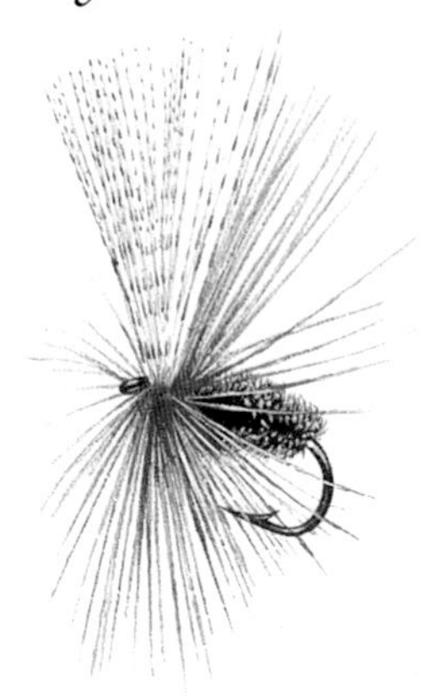

John Storey was keeper on the River Rye in Yorkshire, a post now held by his grandson Arthur. The fly created by John Storey can be classed as a fancy dry fly, imitating nothing in real life except, perhaps, some form of terrestrial beetle, or even a species of Hymenoptera. Whatever it represents, it is still a popular pattern in the North of England.

Hook 12-16.
Thread Black.
Tail None.
Body Peacock herl.
Rib None.
Hackle Natural Rhode Island Red.
Wing Grey mallard tied forward of the hook.

Imperial

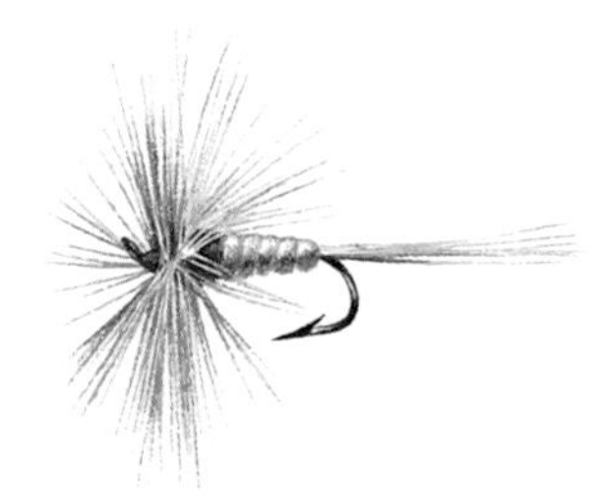

Sometimes called the Kite's Imperial after its originator, the late Major Oliver Kite. The fly is very similar to such flies as the Usk Nailer. It is used as an imitation of the dark olive (*Baetis rhodani*). Oliver Kite died in 1968 whilst on a fishing trip.

Hook 14-16.
Thread Purple.
Tail Greyish brown in early season; later, honey cock hackle fibres.
Body Heron primary herl, doubled to form a thorax.
Rib Fine gold wire.
Hackle Honey dun cock.
Wing None.

Pheasant Tail

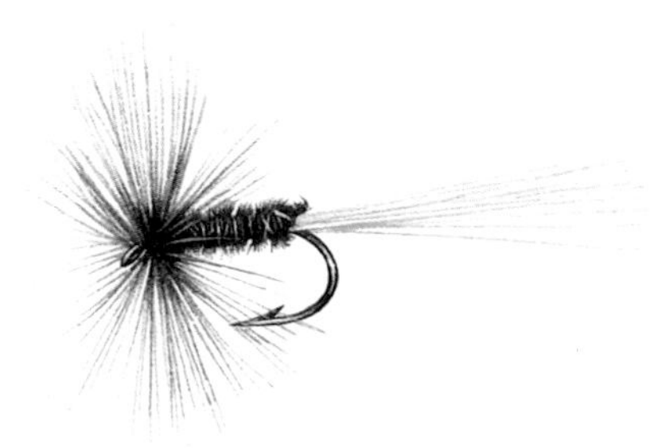

The chestnut brown herls from the centre tail of a cock pheasant provide the fly-tyer with a material whose keynote is versatility. Used for both nymphs and dry flies, it is a fly-dressing medium I could not be without. The dry Pheasant Tail is a pattern from Devon (UK), where it is still considered to be one of the finest. The dressing given was one of Skue's patterns.

Hook 12-16.
Thread Black, or sometimes yellow.
Tail Cock pheasant tail fibres or honey dun cock fibres.
Body Dark red cock pheasant tail fibres.
Rib Gold wire.
Hackle Honey dun cock.
Wing None.

Beacon Beige

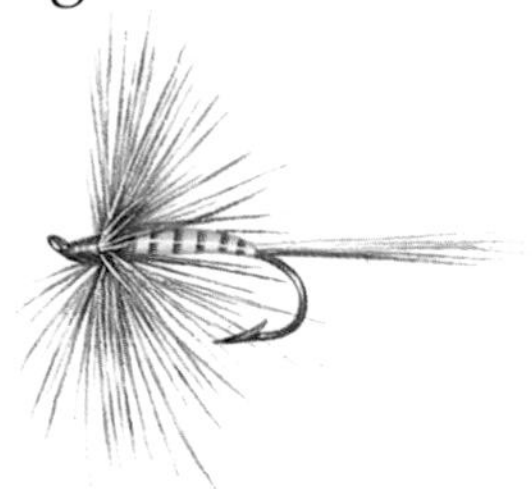

This is another fly, based on an earlier Devon fly, from the vice of Peter Deane. He named this pattern after the hill, Culmstock Beacon, which dominates the valley of the River Culm. This is a good fly with excellent floating properties and a high degree of visibility.

Hook 14-16.
Thread Brown.
Tail Grizzle hackle fibres.
Body Stripped peacock quill from the eye of the feather.
Rib None.
Hackle Grizzle and natural red game.
Wing None.

Terry's Terror

This dry fly was conjured up by Ernest Lock of Andover and his angling friend Dr C. Terry of Bath. It is a good fish taker when olives are on the water, although it certainly does not look anything like the natural olive. It is just one of those flies that work.

Hook 10-16.
Thread Black or brown.
Tail Equal parts of yellow and orange goat hair, clipped short and flared.
Body Peacock herl.
Rib Fine copper tinsel.
Hackle Natural red cock.
Wing None.

Goddard's Caddis

Sometimes called the G & H Sedge, this fly was devised by John Goddard and Cliff Henry at Bough Beech reservoir in Kent, where they both fished. This fly can imitate a wide number of the larger sedge flies and is unique in its wing construction, using buoyant deer hair clipped to a wing shape. This pattern has achieved wide popularity on both sides of the Atlantic.

Hook Long shank 12-14.
Thread Black.
Tail None.
Body Underbody green or yellow wool.
Rib None.
Hackle Two light red cock hackles clipped at the top, stalks left uncut to form antennae.
Wing Spun deer hair cut to shape.

Light Cahill

One of the most popular of the standard American dry flies. It is over one hundred years old and still going strong. This is an extremely versatile fly and is used as an imitation of many of the lighter-coloured ephemerids, such as *Stenacron canadense* or *Stenonema luteum*.

Hook 12-16.
Thread Yellow.
Tail Cream hackle fibres.
Body Cream seal's fur or synthetic dubbing.
Rib None.
Hackle Ginger cock, or cream as alternative.
Wing Wood duck flank.

Dark Cahill

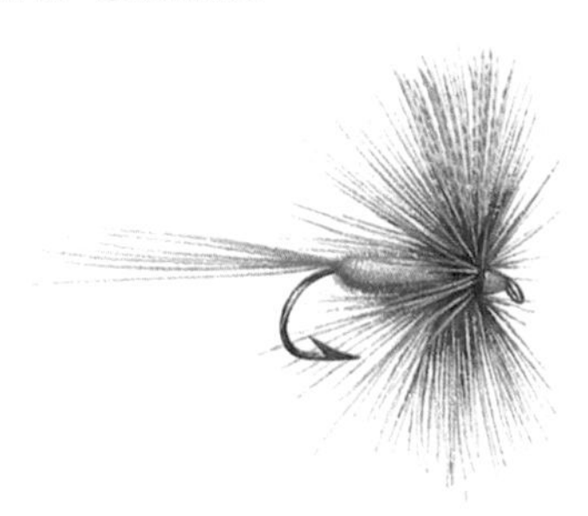

A darker version of the previous fly. Like its sister pattern, it can imitate a wide number of indigenous mayflies, the *Hexagenia atrocaudata*, which hatches from many of the eastern rivers of the USA being just one of them.

Hook 12-16.
Thread Brown.
Tail Brown hackle fibres.
Body Fine brown fur.
Rib None.
Hackle Medium red cock.
Wing Wood duck.

Tup's Indispensable

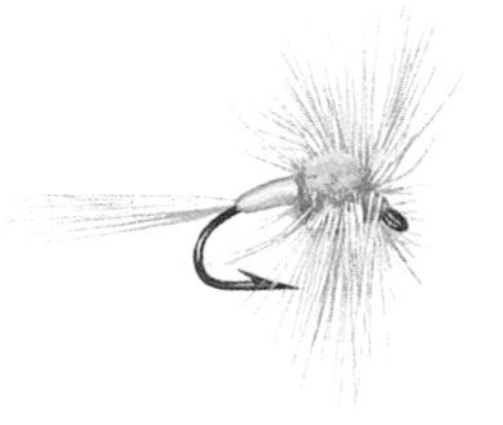

Tup is a country name for a ram and this fly was so named because one of the ingredients used in its construction came from the hair on the scrotum of a ram. The originator of this fly was R. S. Austin, the Devonshire fly-dresser. It was christened by Skues, who rated the fly highly. The thorax was made up from wool from the ever-so-private parts of the ram, cream seal's fur, lemon spaniel's hair, and also some crimson seal's fur. The last-mentioned ingredient was added later, on the advice of Skues, giving a pinkish tinge to the fly.

I have found this fly to be successful when pale wateries are on the river.

Hook 14-16.
Thread Yellow.
Tail Honey dun cock fibres.
Body Two-thirds yellow floss, one-third 'tups' mixture.
Rib None.
Hackle Honey dun.
Wing None.

Light Hendrikson

Another famous standard American dry fly pattern. It is intended to represent the mayfly *Ephemerella subvaria*, or at least the female of the species. A Red Quill is recommended for the male.

Hook 12-14.
Thread Brown.
Tail Medium dun hackle fibres.
Body Pale pinkish-brown fox fur or synthetic substitute.
Rib None.
Hackle Medium dun.
Wing Wood duck fibres.

Quill Gordon

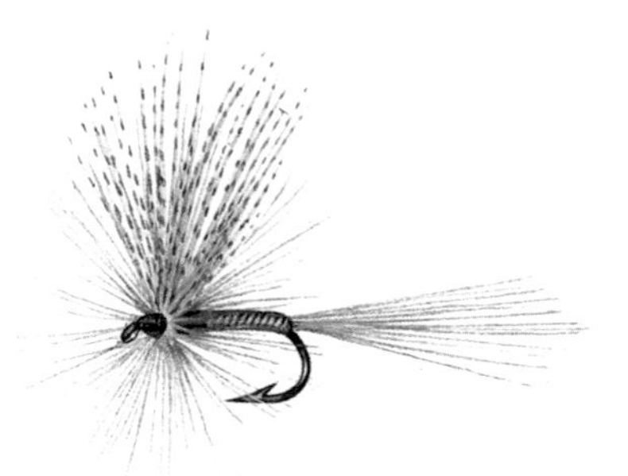

A fly named after the father of American dry-fly fishing, Theodore Gordon. It was first used in the 1890s on the streams of the Catskills. Gordon corresponded regularly with his counterparts in the UK, amongst them Skues.

Hook 12-18.
Thread Cream or light brown.
Tail Medium dun cock hackle fibres.
Body Stripped peacock quill.
Rib Fine gold wire (optional).
Hackle Medium dun cock.
Wing Wood duck.

Dark Hendrikson

As its name suggests, this is a darker version of the previous fly. The Hendrikson was originated around 1915 by Ron Steenrod, who called the pattern Hendrikson after one of his best customers. The Dark Hendrikson came after the light one.

Hook 12-18.
Thread Grey.
Tail Dark dun cock hackle fibres.
Body Dark muskrat or synthetic substitute.
Rib None.
Hackle Dark dun.
Wing Wood duck.

Adams

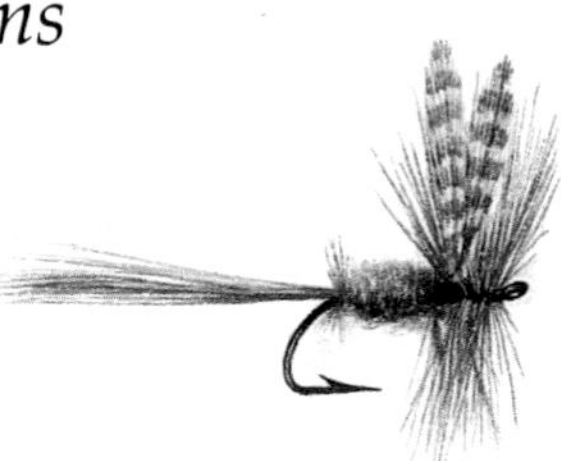

A favourite pattern of mine. I remember a day on the River Itchen when I made short work of about twelve brown trout that took my size 16 Adams for a medium olive. Since that time I have always included this American classic fly in my fly-box. The fly originated from Michigan in the early 1920s.

Hook 10-20.
Thread Grey or black.
Tail Mixed brown and grizzle hackle fibres.
Body Muskrat.
Rib None.
Hackle Mixed brown and grizzle.
Wing Grizzle hackle tips.

Brown Bi-visible

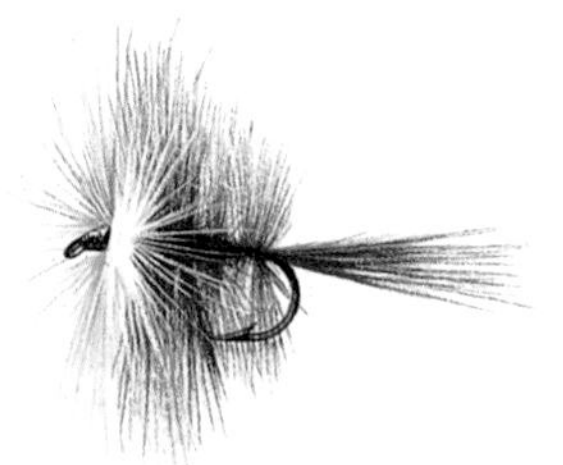

The American Bi-visible series of flies are noted for their good floating properties and also for their visibility to both angler and fish. They are particularly useful on streams and rivers with a good rough and tumble flow.

There are a number of different colours in the series, two of which are illustrated, the Brown and the Grey. On some waters the Black Bi-visible is very killing. The recognition factor is the front white hackle, which is common to all.

Hook 10-14.
Thread Black.
Tail Brown hackle fibres.
Body None.
Rib None.
Hackle Two: palmered brown hackle; white hackle at head.
Wing None.

Grey Bi-visible

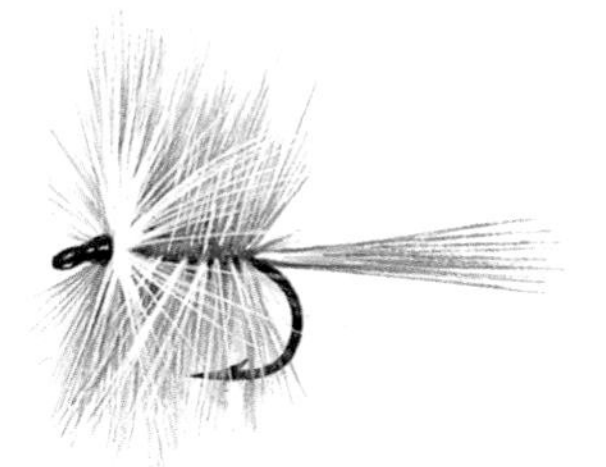

This fly can be very useful as a moth imitation at dusk. In larger sizes all Bi-visibles make excellent dapping flies for trout and seatrout. They are also used as dry flies for salmon.

Hook 10-14.
Thread Grey or black.
Tail Grey cock hackle fibres.
Body None.
Rib None.
Hackle Two: grey palmer; white at the head.
Wing None.

Humpy

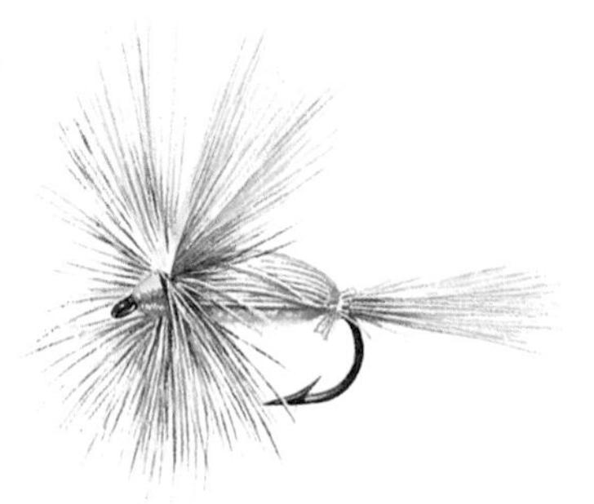

This fly from the fast-flowing rivers of the Western States, USA, is sometimes called the Goofus Bug. A number of body colours are used, especially yellow, red, olive and orange. The underbody of the original Humpy was made from the tying thread. More modern ties tend to use a synthetic fur or antron dubbing.

Hook 10-14
Thread Yellow.
Tail Moose body hair.
Body Natural deer hair over an abdomen or dubbed synthetic light yellow fur.
Rib None.
Hackle Grizzle.
Wing Tips of the deer hair that was used as the body medium.

Rat-faced McDougal

An excellent high-floating American fly. The deer hair body renders the fly unsinkable. In larger sizes it has often been used as a dry fly for salmon.

Hook 10-14.
Thread White.
Tail Dark ginger hackle fibres.
Body Clipped deer hair.
Rib None.
Hackle Dark ginger.
Wing Ginger variant (cree) hackle tips.

White Wulff

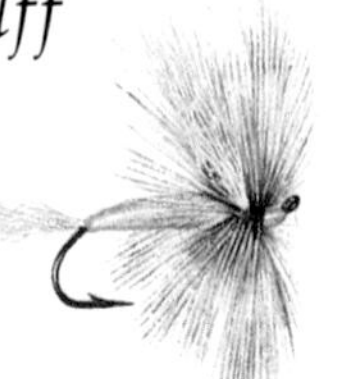

This fly is one of a series created by Lee Wulff, the doyen of American fly-fishing. Though comparatively young by fly standards, these Wulff patterns have quickly become classics. They are fished on both sides of the Atlantic.

Hook 10-14.
Thread White.
Tail White calf tail.
Body White fur or synthetic substitute.
Rib None.
Hackle Badger.
Wing White calf tail.

Royal Wulff

This fly is based on the well-known standard pattern, the Royal Coachman, with a peacock herl and red silk body. Like most of the flies in the Wulff series, it is sometimes used for salmon.

Hook 8-14.
Thread Black.
Tail Natural deer hair.
Body Peacock herl with red floss centre.
Rib None.
Hackle Brown.
Wing White calf tail.

Grey Wulff

This pattern is tied in the parachute style (the hackle is wound around the wing root and not around the shank, as in the more conventional way). Many believe that this style of hackling allows the fly to float more realistically on the water.

Hook 8-14.
Thread Black.
Tail Natural deer hair.
Body Muskrat or synthetic substitute.
Rib None.
Hackle Medium dun (tied parachute, but can also be tied conventionally).
Wing Natural deer hair tips.

King River Caddis

A traditional-style caddis fly originated by American fly-tyer, Bus Buszek. Though modern styles of caddis patterns are now freely available, this conventional dry sedge still maintains a high degree of popularity.

Hook 10-16.
Thread Black.
Tail None.
Body Light brown synthetic fur.
Rib None.
Hackle Brown.
Wing Mottled turkey.

White Irresistable

Who could resist this fly? It is an American pattern for fast-flowing rivers, and is particularly effective at dusk when, of course, the white shows up really well.

Hook 8-14.
Thread White.
Tail White cock hackle fibres.
Body Clipped white deer body hair.
Rib None.
Hackle Badger or white cock.
Wing Badger hackle tips set upright; white can also be used.

Delta Caddis

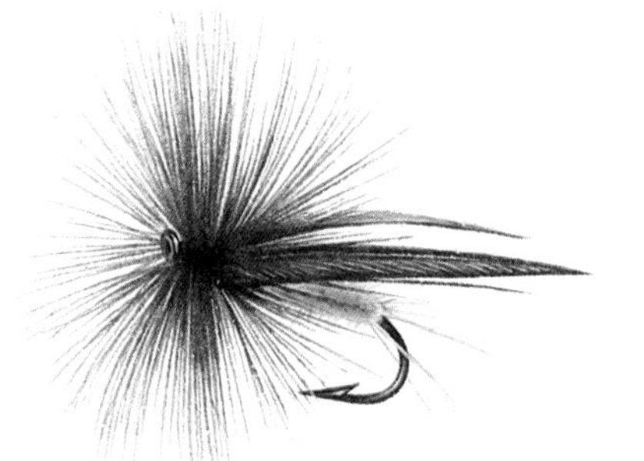

Created by Larry Soloman, this adult caddis imitation copies a dead, or spent caddis lying exhausted in the surface film, after ovipositing. There are a number of colour combinations for the Delta Caddis. Another popular variation has an olive green body and grey hackle tip wings.

Hook 10-14.
Thread Brown.
Tail None.
Body Olive synthetic dubbing.
Rib None.
Hackle Brown.
Wing Brown hackle tips set at 45° to hook shank.

Elk Hair Caddis

The use of deer or elk body hair as a winging medium for caddis fly imitations is now well established on both sides of the Atlantic. Terry Thomas in the UK created a fly a number of years ago which he called the Dark Sedge. This used deer body hair for the wing. The pattern given here was devised by Al Troth. It is used in the American North-West with a great deal of success.

Hook 10-14.
Thread Brown.
Tail None.
Body Olive synthetic dubbing.
Rib None.
Hackle Brown palmered cock hackle.
Wing Natural elk hair. The butts extend to form the head.

Henryville Special

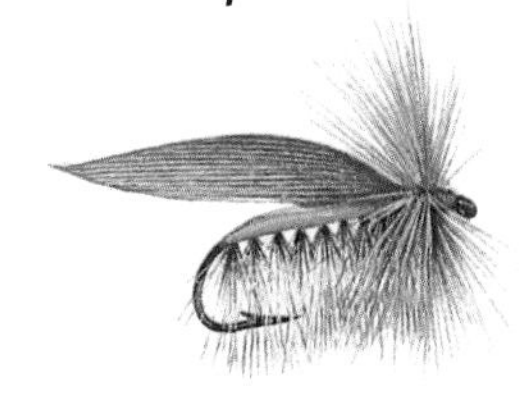

This fly hails from Pennsylvania, USA.

It is based on an earlier British fly similar to the Kimbridge Sedge used on the River Test. This fly goes back to around 1920.

Hook 12-16.
Thread Olive or brown.
Tail None.
Body Light olive floss.
Rib Palmered grizzle cock.
Hackle Dark ginger.
Wing Wood duck flanked by two strips of grey duck wing.

Joe's Hopper

One of the many American patterns devised to imitate the terrestrial grass-hopper. This fly is the fore-runner of such flies as the Whitlock Hopper, which is very similar but utilizes the clipped deer hair, 'muddler'-style head.

I have used this fly as a large caddis imitation on British reservoirs.

Hook Long shank 14-18.
Thread Black or brown.
Tail Red hackle fibres plus a loop of yellow wool.
Body Yellow wool.
Rib Palmered natural red cock, clipped short.
Hackle Natural red cock.
Wing Two strips of mottled turkey wing flanking the body.

Dean's Grasshopper

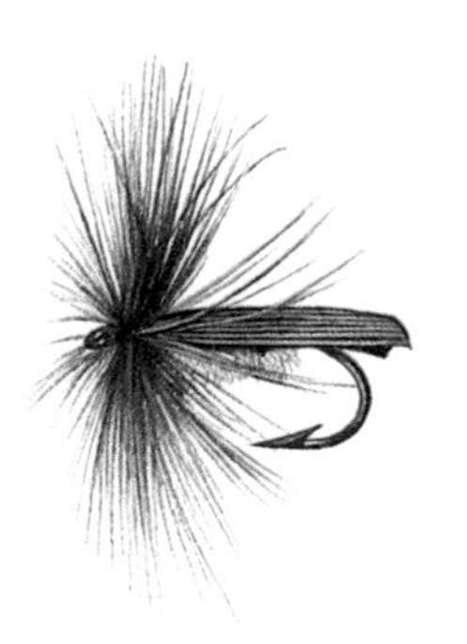

A New Zealand pattern of long standing. The Grasshopper has figured in the fly wallets of fishermen since the days of Charles Cotton. Live grasshoppers are sometimes used as dapping flies on some of the Irish lakes.

Hook 10-12.
Thread Brown.
Tail None.
Body Straw-coloured cock hackles wound palmer style and clipped.
Rib None.
Hackle Brown cock hackle.
Wing Black feather, varnished and clipped in half moon.

Orange Spinner

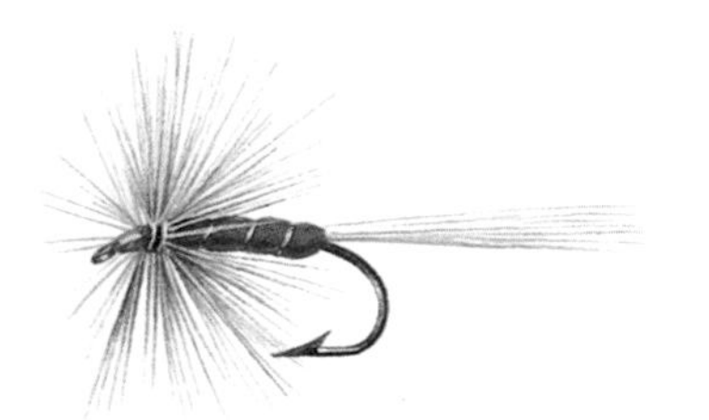

A river dry fly for the summer months. Keith Draper, the renowned New Zealand fly-dresser, thinks it could be taken for an indigenous species of Hymenoptera.

Hook 10-14.
Thread Orange
Tail Ginger cock hackle fibres.
Body Orange wool.
Rib Fine gold wire.
Hackle Ginger cock.
Wing None.

Red Fox

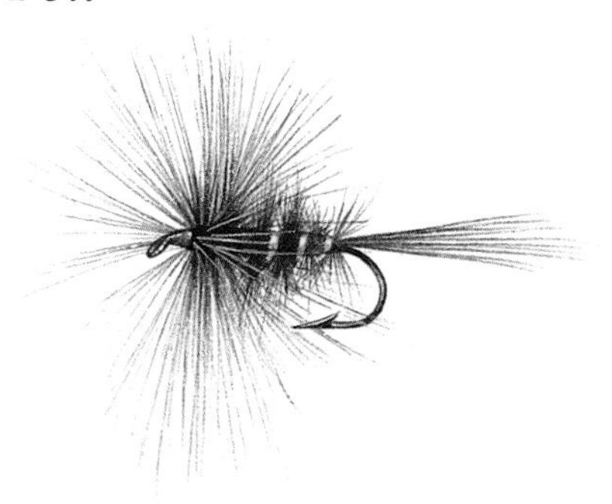

Created by the New Zealand fly-tyer R. K. Bragg, this is very similar in appearance to the British Blue Dun, and like its UK counterpart it is supposed to represent a similar natural ephemerid.

Hook 14-16.
Thread Yellow.
Tail Light natural red cock.
Body Blue squirrel fur mixed with hare's ear.
Rib Fine flat gold tinsel.
Hackle Light red with blue dun in front.
Wing None.

Shoo Fly

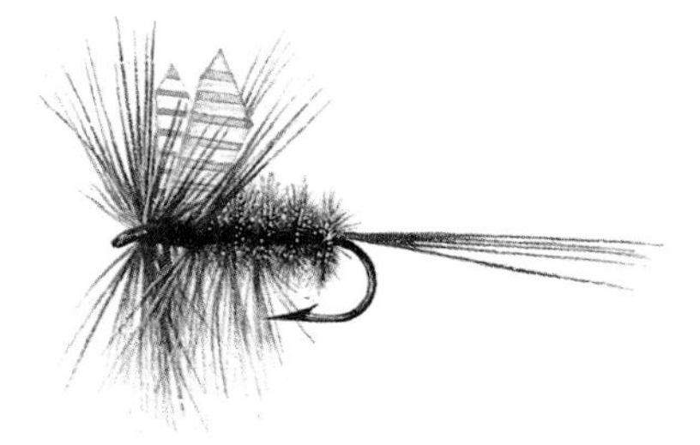

A fancy pattern from New Zealand, for use on stillwaters. The original winging medium called for a veined plastic. The pattern shown here uses raffene (Swiss straw).

Hook 10-14.
Thread Black.
Tail Scarlet cock.
Body Peacock herl.
Rib None.
Hackle Rhode Island Red hackle.
Wing Veined plastic artificial wing.

Brandon's Claret

This is a fancy dry fly from Australia. The three cock hackles at the head ensure that this fly is, at the very least, a good floater.

Hook 10-12.
Thread Black.
Tail Golden pheasant tippet fibres.
Body Fat claret seal's fur or mohair.
Rib Oval gold.
Hackle Three black cock hackles.
Wing None.

Red Bug

An Australian version of the Red Tag. In fact this fly is almost a direct throwback to the original fly fished in Macedonia.

Hook 10-12.
Thread Black.
Tail Red wool.
Body Red wool.
Rib Oval gold tinsel.
Hackle Stiff cock hackle over one-third of the hook.
Wing None.

Tri-tree Beetle

This is a typical terrestrial pattern imitating one of the indigenous Australian beetles. Every fly-fishing country has its own beetle patterns, and Australia is no exception. This pattern can be fished wet or dry.

Hook 10-12.
Thread Black.
Tail None.
Body Black ostrich herl.
Rib None.
Hackle Black cock hackle.
Wing case Brown feather tied over the back.

Mooi Moth

Most South African dry flies are derived from, or are actual, British flies. This pattern, the Mooi Moth, is pure-bred South African.

Hook 10-12.
Thread Black.
Tail Medium blue dun cock hackle fibres.
Body Stripped peacock quill from the eye.
Rib None.
Hackle Medium blue dun.
Wing Grey duck.

Grizzle Hackle

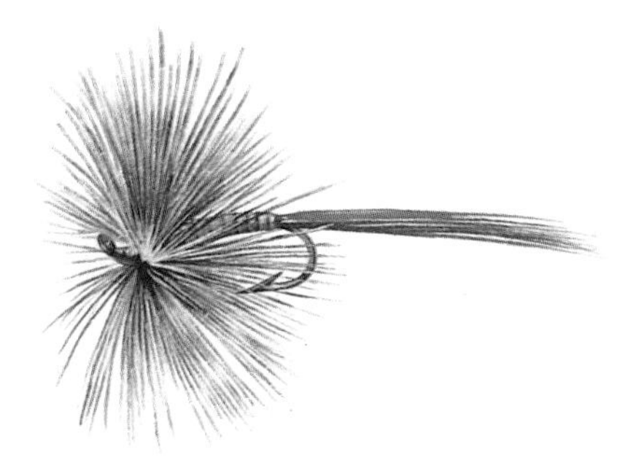

A high-floating fly from South Africa that is suitable for the streams and rivers that tumble out of the Drakensburg mountains. Most fishing in South Africa is done with the wet fly, but a number of anglers find the dry fly useful on certain stretches of rivers and also on the dams, which are to be found all over the republic, some stocked with trout, others with large-mouth bass.

Hook 10-14.
Thread Black.
Tail Natural red cock hackle fibres.
Body Stripped peacock quill from the eye.
Rib None.
Hackle Two grizzle cock hackles.
Wing None.

Diabolo

This is a French pattern for rough water. The fore-and-aft hackle helps its floating potential. This fly is the creation of the Guy Plas of Marcilac-la-Croisille. There are a number of colour variations for the body: yellow, olive, grey, and the red version depicted here.

Hook 12-16.
Thread Black or red.
Tail None.
Body Red silk, varnished.
Rib None.
Hackle Blue dun (dark) fore-and-aft.
Wing None.

Tricolore

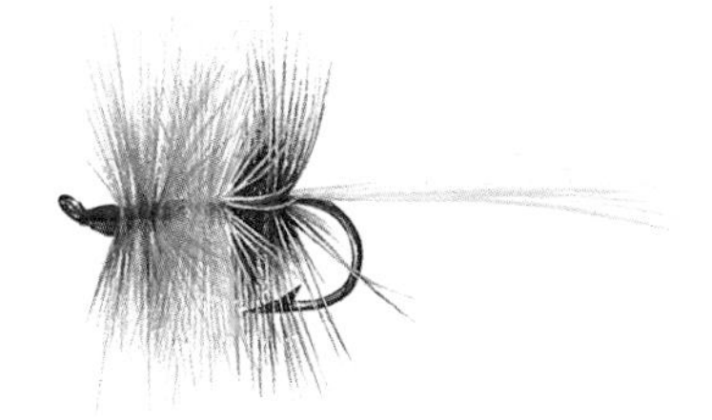

A French palmered fly similar to the Bi-visible series of dry flies. This pattern is used for both trout and grayling throughout Europe.

Hook 12-14.
Thread Black.
Tail Light ginger cock hackle fibres.
Body None.
Rib None.
Hackle Three: black, ginger, and blue dun. (There are other colour combinations.)
Wing None.

Fourmi Brun

A favourite fly of the late Charles Ritz, one of France's greatest fly fishermen as well as a leading hotelier. The time to use the ant in Europe is late summer.

Hook 12-16.
Thread Black or brown.
Tail None.
Body Peacock herl/orange floss/cock pheasant tail fibres.
Rib None.
Hackle Natural red cock.
Wing Blue dun hackle tips set over the back.

Pont Audemer

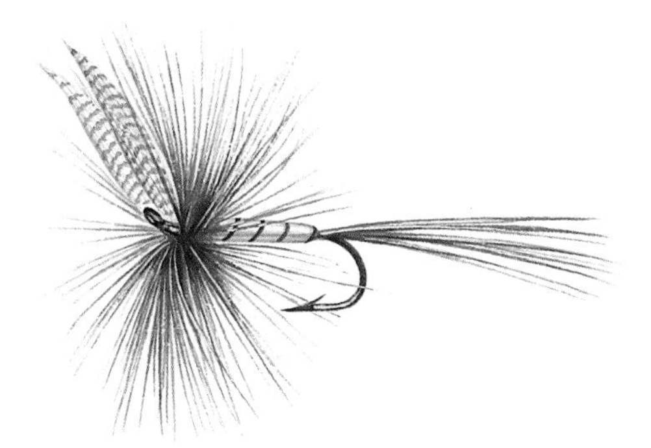

A traditional fly from Normandy. There are a number of variations of this fly. All are used from May onwards on the chalkstreams of that area.

Hook 10-14.
Thread Black.
Tail Red cock hackle fibres, or cock pheasant tail fibres.
Body Natural raffia.
Rib Black silk.
Hackle Two medium natural red.
Wing Grey speckled duck tied over the eye.

Favourite de Carrére

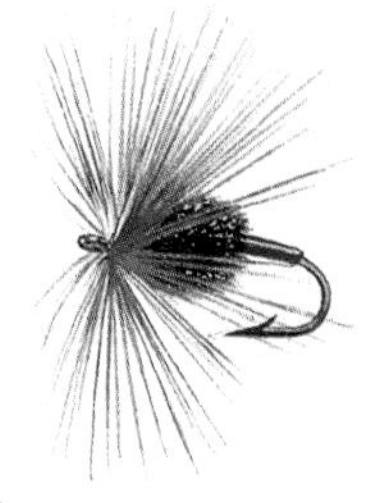

This fly could be described as a French version of the Red Tag. It is used for both trout and grayling and can be fished wet or dry.

Hook 10-14.
Thread *Black.*
Tail None.
Body Peacock herl with a red silk tip.
Rib None.
Hackle Natural red.
Wing None.

Diaphane

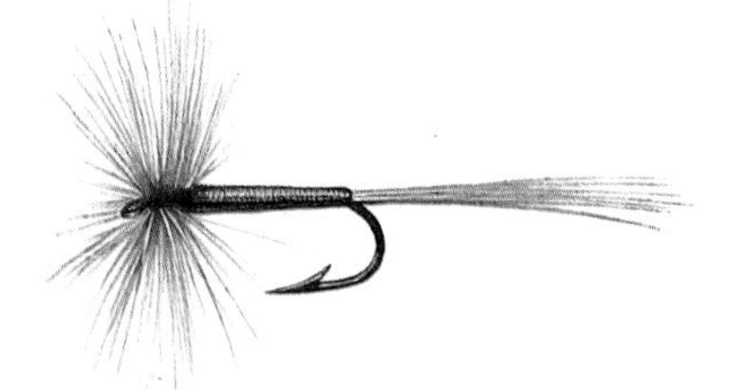

Another pattern from the Guy Plas stable. The hackle on this fly should slope slightly towards the eye. The Diaphane is recommended for those difficult fish found in the smooth glides of a stream.

Hook 14-18.
Thread Olive.
Tail Blue grey.
Body Thin olive silk, varnished.
Rib None.
Hackle Blueish-grey dun.
Wing None.

L'Olive Moyenne

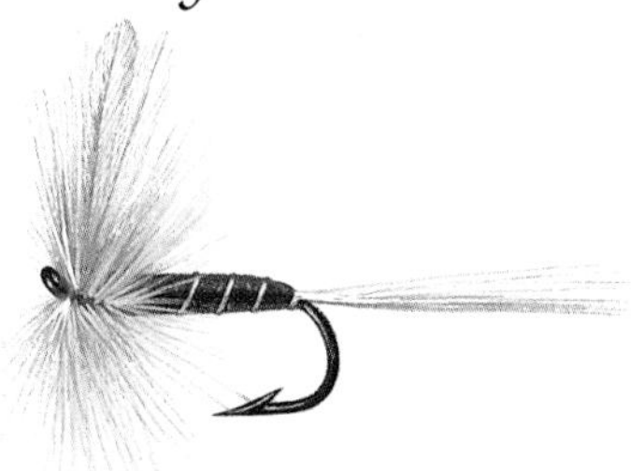

This pattern is one of a series of flies known as the Gallica series – I believe this one to be number 9. It is an imitation of the medium olive. The original dressing called for a dyed natural raffia body. A PVC material has been used for the example shown here.

Hook 12-14.
Thread Olive.
Tail Blue dun hackle fibres.
Body Thin brown PVC.
Rib Yellow silk.
Hackle Brownish olive with blue dun in front.
Wing Two blue dun hackle tips with a mauvish tinge.

Panama

A popular pattern in France. In the past years I have been asked to tie quite a number of these flies for British anglers. The fly can best be described as a fancy dry pattern.

Hook 10-14.
Thread Black.
Tail Golden pheasant tippett.
Body Natural raffia tipped with four turns of black silk.
Rib Palmered (natural) red cock.
Hackle Natural red cock; front hackle brown partridge.
Wing Cree hackle points.

Roman Moser Dun

One of the most realistic dry olive dun patterns I have ever seen, this fly, tied by one of Austria's leading fly-dressers, uses up-to-the-minute materials. The interesting feature of this fly is the wing, which is pre-formed.

Most of the Roman Moser's flies use man-made products in their construction, available from the fly-dressing house, Traun River Products.

Hook 12-14.
Thread Yellow.
Tail Light ginger cock hackle fibres.
Body Yellow or olive poly dubbing.
Rib None.
Hackle Light ginger, palmered down the body, then clipped underneath.
Wing Pre-formed wing manufactured for this purpose.

Roman Moser Sedge

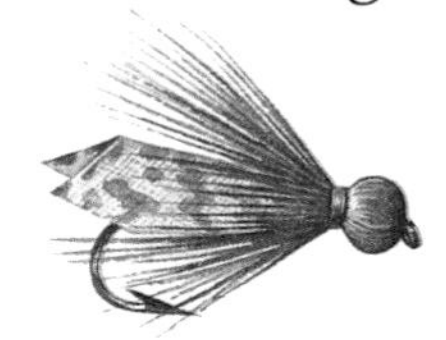

Like many modern caddis/sedge patterns, this one from the Moser stable also uses the good floating properties of deer body hair. In this fly it is used for the body, legs and head.

Hook 12-8.
Thread Brown.
Tail None.
Body Deer hair bound lengthways along the hook.
Rib None.
Hackle Collar of deer hair.
Wing Pre-formed sedge wing.
Head Deer hair.

Roman Moser Stonefly

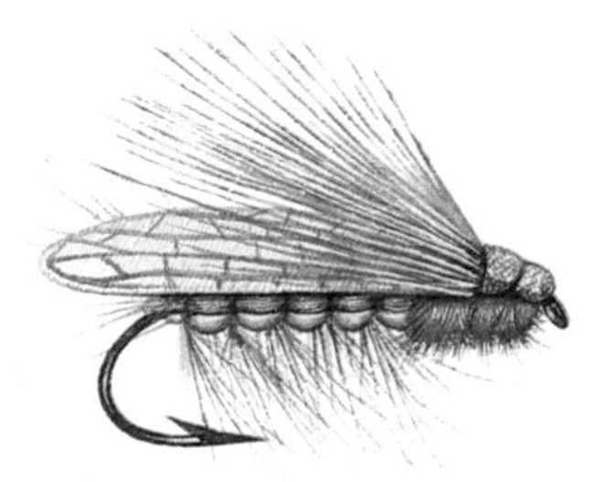

Like his dun pattern, this stonefly imitation also uses modern synthetic materials to achieve a most realistic fly. Roman Moser's flies are almost too good to fish with.

Hook Long shank 14-8.
Thread Brown.
Tail None.
Body Thin foam plastic, yellow colour.
Rib None.
Hackle Palmered blue dun, with a deer hair collar to simulate legs.
Wing Special pre-formed stonefly wing.
Head Thin grey plastic foam.

Victor's Sedge

I obtained this fly from Victor Salt of Madrid, who based his pattern on a French type of fly. The hackles come from the fabled cocks of Leon, world-renowned for the quality of their shiny, stiff feathers.

Hook 14-10.
Thread Brown.
Tail None.
Body Cock pheasant tail fibres.
Rib None.
Hackle Natural red cock hackle.
Wing Mottled fibres from the spade hackle (Flor de Escoba).

Mayfly

The natural mayfly occurs on many of the slower-running rivers of Austria during May and June. This pattern uses a pre-formed detached latex body, which not only looks realistic, but also helps the fly to float well.

Hook 10-12.
Thread Black or brown.
Tail Three cock pheasant tail feathers.
Body Pre-formed detached latex.
Rib None.
Hackle Olive cock.
Wing Mallard breast feathers dyed either yellow, olive or natural.

N. E. Sedge

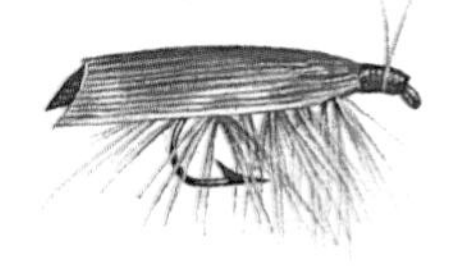

This typical Austrian sedge was devised by Norbert Eipeltauer of Vienna. Sedge flies are used on many of the Austrian mountain streams.

Hook 10-14.
Thread Black.
Tail None.
Body Brown or olive nylon dubbing.
Rib None.
Hackle Palmered brown cock hackle.
Wing Hen pheasant.

White Moth

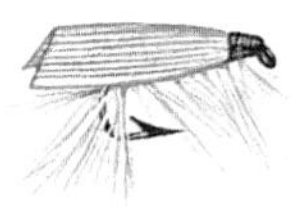

This Austrian pattern is sometimes called the White Sedge. It is a good fly for both trout and grayling, especially at dusk.

Hook 12-16.
Thread Black.
Tail None.
Body White nylon or poly dubbing.
Rib None.
Hackle Palmered white cock.
Wing White duck or goose.

Grey Spinner

This Austrian fly is used for both trout and grayling. In some Austrian rivers grayling are found up to 2.27kg (5 lb). Small grey flies are very popular in many parts of Europe.

Hook 14-16.
Thread Black or grey.
Tail Five to eight stiff grey cock hackles.
Body Grey poly dubbing.
Rib None.
Hackle Natural blue/grey dun.
Wing Two hen hackle points.

Chochin

This small fly from Luis Antunez Jnr, from Madrid, represents a wide number of small Diptera species that get caught in the surface film. This is termed a fore-and-aft fly, having a hackle at either end, which enhances the floating potential of the fly.

Hook 14-16.
Thread Olive.
Tail None.
Body Flat gold tinsel with clear green swannundaze over.
Thorax Brown synthetic fur.
Rib None.
Hackle Tail hackle pale blue dun; head hackle dark blue dun.
Wing None.

Verano Amarillo

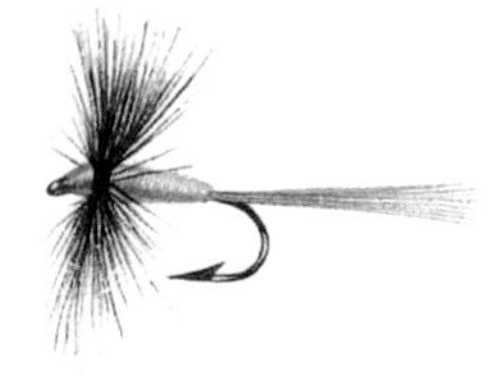

The Yellow Summer is another fly created by Luis Antunez, a young fly-dresser of great skill and ingenuity. This fly imitates one of the summer olives.

Hook 12-14.
Thread Primrose.
Tail Medium blue dun.
Body Two-thirds yellow goose or swan fibre, one-third green fibre.
Rib None.
Hackle Medium blue dun.
Wing None.

Hoz Seca

Another ephemerid-imitating pattern from Luis Antunez. *Seca*, of course, means dry; but as for *Hoz*, it can mean a sickle or a ravine, but in this case it is the name of a river.

Hook 14.
Thread Light brown.
Tail Medium blue dun.
Body Dirty olive synthetic fur.
Rib Close turns of fine copper wire.
Hackle Blue dun and light red, mixed.
Wing Dirty grey polypropylene.

Tajo

A dry fly named after the famous river of Spain and Portugal, the Tagus, which flows into the sea at Lisbon. Like the other Spanish flies, Hoz Seca and Verano Amarillo, this is a mayfly representation.

Hook 14-16.
Thread Primrose.
Tail Medium blue dun.
Body Varnished orange silk with a tip of dark grey.
Rib None.
Hackle Blue dun.
Wing Grey polypropylene.

Moustique flies

The Moustique flies hail from the Jura region of Switzerland, which borders on France, and have been used for over one hundred years. The most interesting feature is the hackle. The three flies shown here use the delicate grey feather situated near the preen gland of a duck. This feather provides good floatation, as one would expect, because it is coated with the oil that the duck uses to waterproof its feathers. See also M. Fratnik's F. Fly from Slovenia, a simple pattern which uses the gland feathers in a slightly different way.

Moustique 1

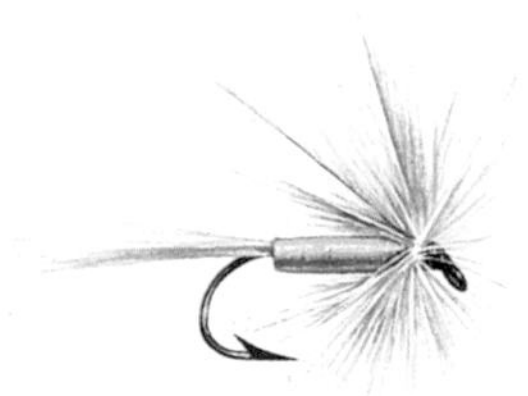

This fly from Switzerland is a representation of some of the smaller, pale-coloured ephemerids, the equivalent perhaps of the British species, the pale watery (*Baetis bioculatus*) or even the small spurwing (*Centroptilum luteolum*).

Hook 14-16.
Thread Black or brown.
Tail Medium-blue dun hackle fibres.
Body Yellow floss silk.
Rib None.
Hackle Feather from a duck's green preen gland, clipped to size.
Wing Two blue dun hackle tips.

Moustique 2

The Swiss equivalent of the well-known Red Spinner fly.

Hook 14-16.
Thread Black.
Tail Ginger hackle fibres.
Body Red floss silk.
Rib None.
Hackle Feather from a duck's preen gland, clipped to size.
Wing None.

Moustique 3

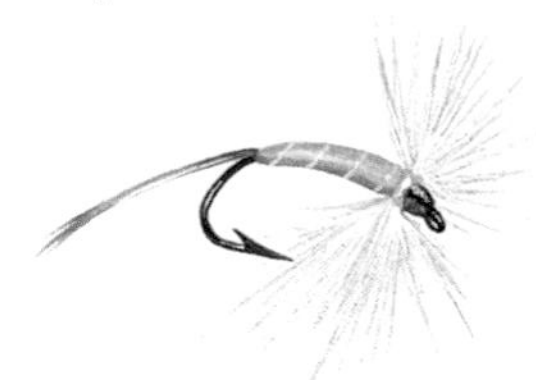

All three Moustique flies are used for both trout and grayling. They are strongly influenced by French flies both in design, and in the material used in their construction.

Hook 14-16.
Thread Black.
Tail Blue dun.
Body Light green floss.
Rib Yellow silk.
Hackle Feather from a duck's preen gland, clipped to size.
Wing None.

Universal Sedge 1 and 2

These two Swiss sedge or caddis patterns can best be described as broad-spectrum flies, used to imitate a wide number of natural sedges. The wings of both patterns are tied flat along the body. The two slips of feather are tied one on top of the other, forming the distinctive 'V' at the rear. Examples of these flies in my collection have lacquered wing to give additional strength to the fly.

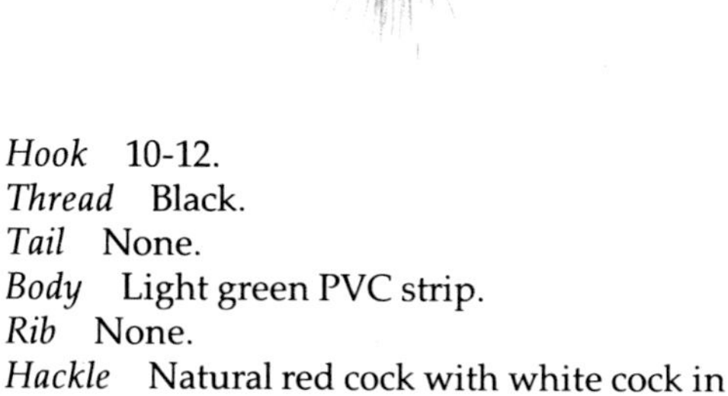

Hook 10-12.
Thread Black.
Tail None.
Body Light green PVC strip.
Rib None.
Hackle Natural red cock with white cock in front.
Wing Hen pheasant wing quill, tied flat on top.

Hook 12-14.
Thread Black.
Tail None.
Body Mustard yellow plastic foam.
Rib Clear nylon thread.
Hackle Dark red game cock hackle.
Wing Hen pheasant wing quill.

The 'F' Fly

This particular pattern was created by Marjan Fratnik of Milan (formally of Most Na Soci in Slovenia). He was aquainted with the Moustique series of flies from the Swiss Jura. However, good as they are, Marjan thought them to be a little fragile, and he developed his 'F' Fly. This unique dry fly, using the feathers from the duck's preen gland, has been tested by fly fishermen all over Europe and found to be a most killing pattern. It is used for both trout and grayling.

Hook 10-18.
Thread Black, grey, olive or yellow.
Tail None.
Body Tying thread or heron herl.
Rib None.
Hackle and wing The hackle and wing are combined: a small duck gland feather.

Slovenica Series

This series of excellent sedge flies was developed by Dr Bozidar Voljc after an extensive study of the sedges of the Slovenian mountain and chalk streams. Perhaps the unique factor in Dr Voljc's flies is their almost indestructible nature, which is due in the main to his method of winging. The wings are prepared by coating the feather with PVC glue and sticking it to non-elastic nylon (such as nylon stockings). When dry, the feather and nylon are cut and trimmed to shape. It is this nylon that gives strength to the fly.

The Silver Sedge

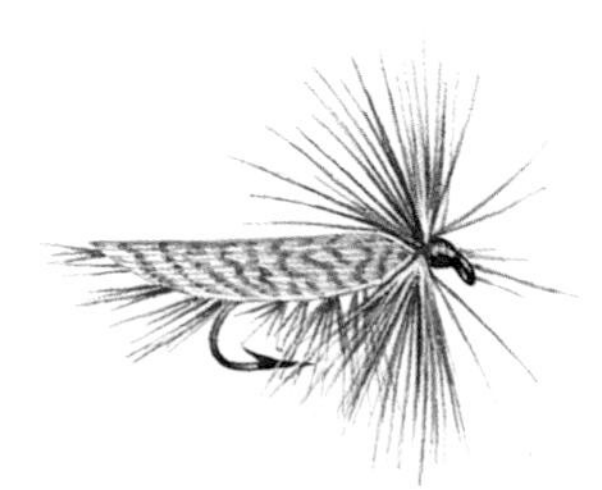

Hook 14-16.
Thread Black.
Tail None.
Body Palmered dark ginger cock.
Rib None.
Hackle Dark ginger.
Wing Grey partidge, glued to non-elastic nylon, then trimmed.

The Grouse Wing

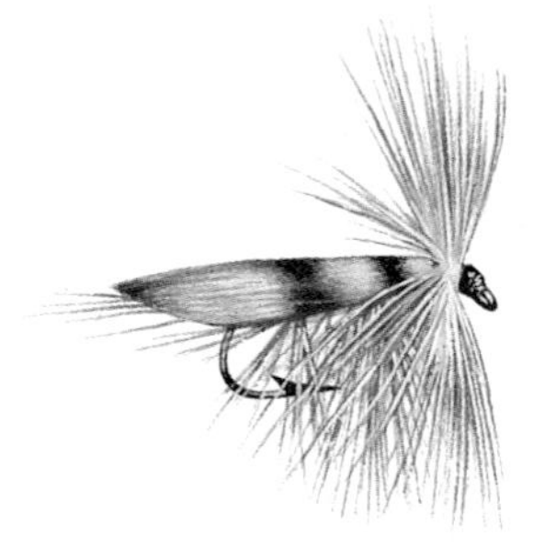

Hook 14.
Thread Black.
Tail None.
Body Palmered dark ginger.
Rib None.
Hackle Dark ginger.
Wing Woodcock body feather glued to nylon.

The Cinnamon Sedge

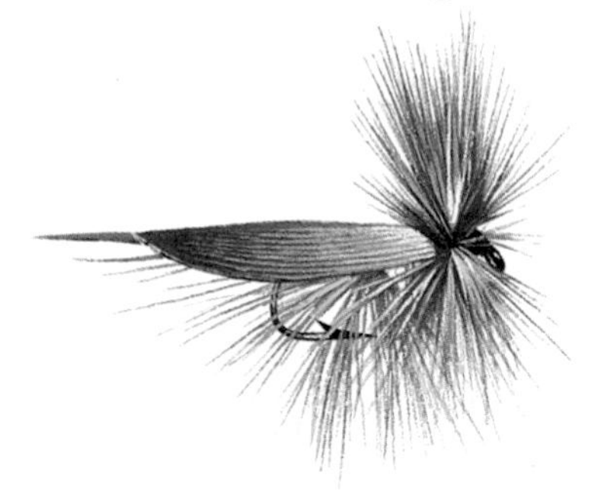

Hook 12-18.
Thread Black.
Tail None.
Body Palmered dark ginger cock.
Rib None.
Hackle Dark ginger.
Wing Hen mallard body feather glued to nylon.

The Grey Sedge

Hook 12-16.
Thread Black.
Tail None.
Body Palmered grizzle hackle.
Rib None.
Hackle Grizzle.
Wing Grey partridge glued to nylon.

Black Sedge

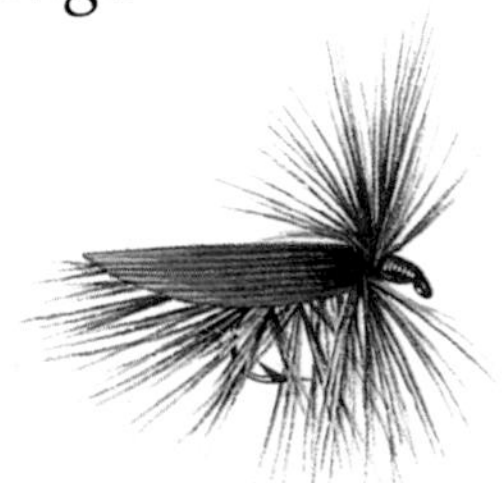

Hook 12-16.
Thread Black.
Tail None.
Body Palmered black cock.
Rib None.
Hackle Black cock.
Wing Black hen body feather glued to nylon.

The Caperer

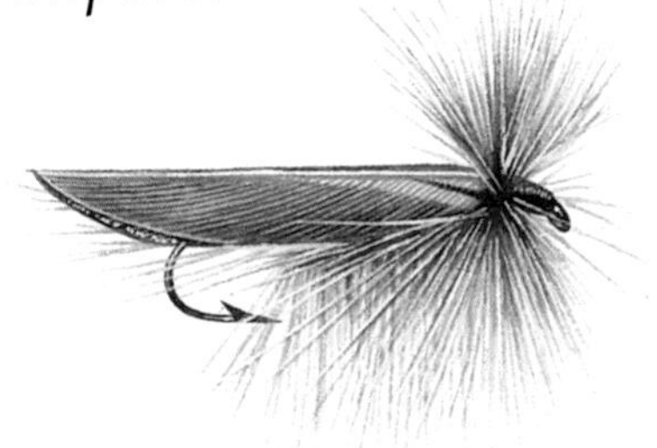

Hook Long shank 8.
Thread Black.
Tail None.
Body Palmered dark cream badger.
Rib None.
Hackle Dark cream badger, or greenwell cock.
Wing Cock pheasant flank glued to nylon.

Carniolica Series

The Alpine rivers of Slovenia are rich in many species of stoneflies. Dr Bozidar Voljc of Ribnica Na Dolenjskem, Slovenia, in his Carniolica flies has sought to produce many of these stoneflies. Dr Voljc is both an acknowledged entomologist and author, as well as being an excellent fly-tyer and fisherman.

Yellow Sally

Hook 14-18.
Thread Yellow.
Tail None.
Body Palmered, dyed yellow cock hackle.
Rib None.
Hackle Yellow cock.
Wing Yellow hackle treated with a PVC adhesive and trimmed to shape.

The Needle Fly

Hook 18.
Thread Black.
Tail None.
Body Palmered dark blue dun.
Rib None.
Hackle Dark blue dun.
Wing Blue dun hackle treated with PVC adhesive.

The Brown Stonefly

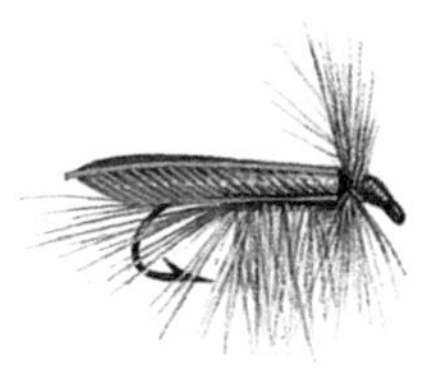

Hook 14-16.
Thread Black.
Tail None.
Body Palmered natural red cock.
Rib None.
Hackle Natural red cock.
Wing Natural red cock hackle treated with PVC.

The Willow Fly

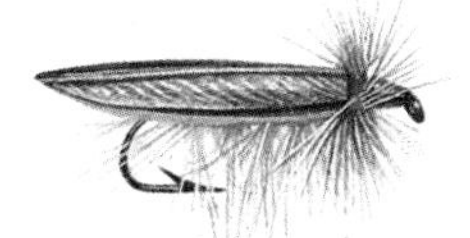

Hook 14.
Thread Black.
Tail None.
Body Palmered medium blue dun.
Rib None.
Hackle Medium blue dun.
Wing Medium blue dun treated with PVC adhesive.

The Large Stonefly

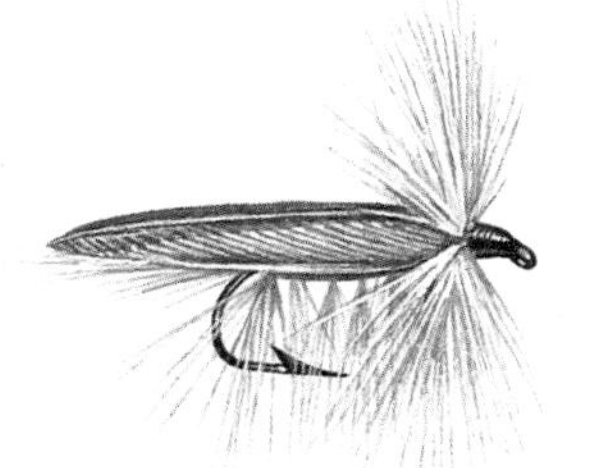

Hook Long shank 12.
Thread Black.
Tail None.
Body Yellow polypropylene, palmered with medium-blue dun cock hackle.
Rib None.
Hackle Medium blue dun.
Wing Medium blue dun treated with PVC adhesive.

Bird's Stonefly

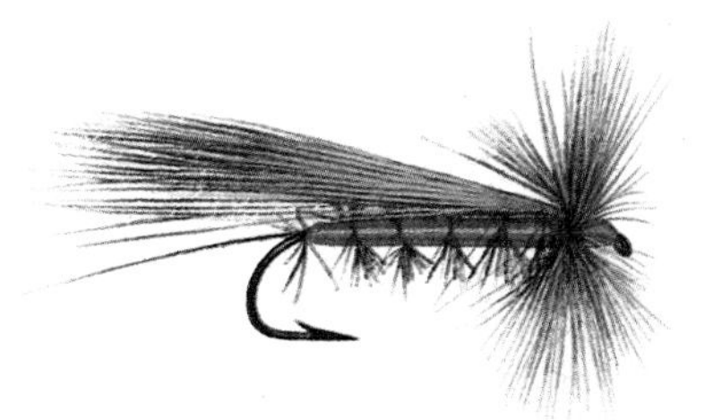

This fly goes back to about 1960 and was the creation of Calvert T. Bird of San Francisco. It was used to imitate the large, natural stoneflies of the Montana streams. Since its inception, it has spread throughout the north-western United States and into Canada. In smaller sizes I use it as a useful dry caddis imitation.

Hook Long shank 4-8.
Thread Orange or brown.
Tail Two strands of black hair from a brown bear's paw.
Body Burnt orange floss.
Rib A palmered furnace hackle clipped short.
Hackle Furnace hackle trimmed and lacquered to project at the sides.
Wing Brown bucktail.

Hornberg

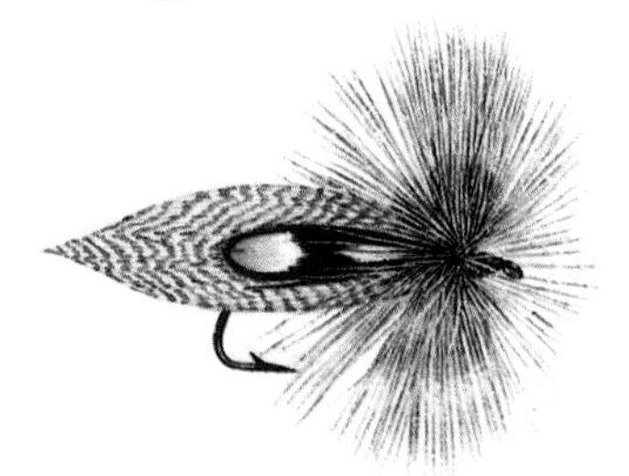

This is a most unusual pattern, sometimes called the Hornberg Special Streamer. It was created by Frank Hornberg, Wisconsin, USA. It is a popular pattern throughout the USA and through into Canada, where it has many adherents in Quebec. It is used as a dry sedge pattern, but it is one of those flies that, when it sinks, is then fished as a wet fly.

Hook Long shank 8-12.
Thread Black.
Tail None.
Body Flat silver tinsel.
Rib None.
Hackle Two grizzle cock.
Wing Yellow cock hackle fibres, flanked by grey mottled mallard flank. The tips of the mallard feather are varnished together.
Cheek Jungle cock (optional).

Renegade

This dry fly can be used on both river and lake. It is considered to be an excellent pattern for cutthroat trout. It was devised in Wyoming by Taylor 'Beartracks' Williams around the late 1920s. Since that time it has been used throughout the Pacific North West right up into Alaska.

Hook 8-14.
Thread Black.
Tip Flat gold tinsel.
Tail None.
Body Peacock herl.
Rib None.
Hackle Rear hackle brown; front hackle white.
Wing None.

WET FLY PATTERNS

The trout finds most of its food beneath the surface of the water, sometimes by grubbing around the weed-beds, at other times by rising in the body of the water to take nymphs and pupae on their way to the surface.

The wet flies depicted in this section fall into various categories. Some represent larval and pupal forms of various aquatic insects; others imitate drowned adults, or even swamped still-born flies. Some wet flies are tied to represent drowned terrestrials, such as beetles. A high proportion imitate nothing in Nature, but are classed as attractor flies; flashy creatures designed to stimulate the fish and tempt them to take out of curiosity, or perhaps even anger. A number of the silver-bodied flies can emulate small fry.

Most of the flies given in the dry-fly section have their wet-fly equivalents.

The use of heavier hooks, softer hen hackles instead of cock, and in the case of winged flies a backward-sloping wing, changes the dry fly into a wet one.

There are two main, traditional areas of wet-fly fishing. Firstly, there are the wild, rough and tumble rivers and streams, the troubled, rain-fed waters, where it is difficult to see a fish rise and certainly as difficult to see a minute dry fly on the surface. On such waters, wet flies are used almost exclusively, fished both upstream and down, as necessity or the terrain dictates. The second main area of wet-fly fising is on lakes, lochs and reservoirs, where the traditional drifting-boat method enables the angler to fish a team of wet flies just below the surface. To this day, this is still one of the most effective methods of taking still-water trout, and is a method insisted upon in international fly-fishing competition rules.

I have to confess that I am seldom successful on wild streams if I use some of the established nymph patterns (patterns, I might add, that I use on chalkstreams quite successfully). Perhaps I do not fish them with confidence or conviction. However, it is a different story when I resort to the traditional, soft-hackled wet flies. The Partridge and Orange, the Snipe and Purple, the Black Spider, a wet Coch-y-Bonddu, and many others, have all taken their fair share of trout for me.

People often ask, 'When do you fish a wet fly, and when a dry?' I always fish a dry fly pattern when I see a trout rising during a hatch of natural insects. The secret is to keep your eyes open, watching the water for the slightest dimpling of the surface. There are times, however, when, though there appears to be surface activity, the trout refuses to succumb to a dry fly, no matter how exact the imitation or perfect the presentation. If this is the case, the remedy can often be found in fishing just below the surface with a wet fly. When no activity is obvious, it is a case for the wet fly, pure and simple.

The soft, game bird hackles of many wet flies have the necessary mobility in the water, they pulsate and 'kick' in the current, giving life to the fly, attracting the fish by their very movement. They look alive, they look edible; the two key properties for a successful fly.

Of all the past authorities on the wet fly, T. E. Pritt, in his book *Yorkshire Trout Flies* (subsequently republished as *North Country Flies*), has had a lasting effect on wet-fly fishing and flies as we know them today. Before Pritt, who published his work in 1885, other writers produced works of excellence on wet flies: John Jackson in his book, *Practical Fly Fisher* (published in 1854); the much quoted W. Blacker and his book, *The Art of Angling and Complete System of Fly Making and Dyeing of Colours* (1843). Then there is Michael Theakston in

British Trout Flies (1853). One of the greatest advocates of spider-dressed flies was W. C. Stewart, author of the *Practical Angler*.

Books on wet flies are still being published on both sides of the Atlantic. Their use, and interest in them, is as strong as ever and old-established patterns and methods are enjoying a new-found popularity.

Wherever the swift waters run and the large lakes hold trout, there will always be a place for the wet fly, and fly-fishermen will avail themselves of such weapons in their quest for the spotted fish.

Partridge and Orange

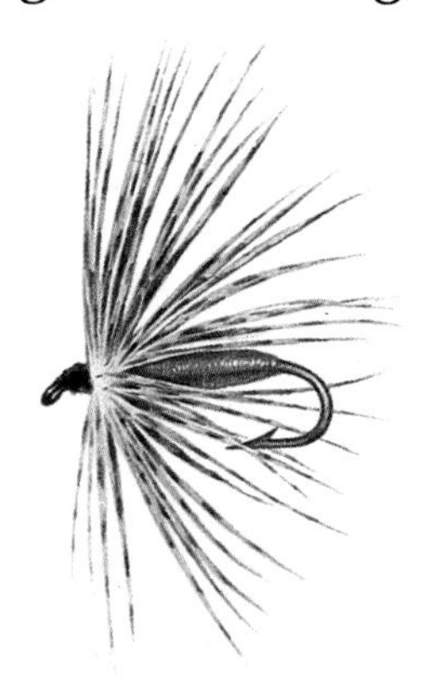

One of the classic English North Country spider-style flies, now used all over the fly-fishing world. It is an effective imitation of many of the early stoneflies (Plecoptera).

Hook 12-16.
Thread Orange.
Tail None.
Body Orange silk.
Rib None.
Hackle Brown partridge.
Wing None.

Partridge and Yellow

A lighter version of the previous fly, this one is found with very little alteration in northern Spain, and also in Italy. It is thought to imitate some of the olives.

Hook 12-16.
Thread Yellow.
Tail None.
Body Yellow silk.
Rib None.
Hackle Light partridge breast.
Wing None.

Snipe and Purple

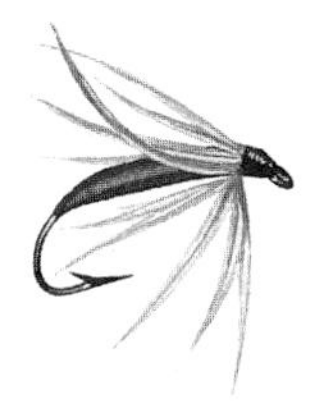

Though fished in many parts of the British Isles, this fly is as Yorkshire as Yorkshire pudding. Ask any fly fisherman from that county what fly he would not be without, and the reply would come back – Snipe and Purple. This fly is a wet version of the Iron Blue (*Baetis niger* or *pumilus*).

Hook 12-16.
Thread Purple.
Tail None.
Body Purple.
Rib None.
Hackle A hackle from the inside of a snipe's wing.
Wing None.

Poult Bloa

The term 'bloa' is from the same root as 'blae', meaning a smokey blueish grey. This spider fly is used as an imitation of many of the ephemerids, and is a wet version of the Pale Watery.

Hook 14-16.
Thread Yellow.
Tail None.
Body Lightly dubbed blue dun fur on yellow silk.
Rib None.
Hackle A hackle from the inside of the young grouse wing.
Wing None.

Butcher

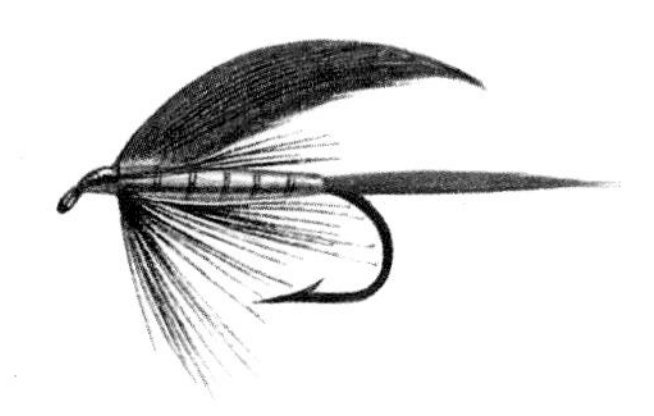

This fly is a long-established attractor pattern. Its history is well documented, being the creation of two gentlemen from Tunbridge Wells, one of whom was a butcher by trade. The colours of the fly supposedly represent the hallmarks of his trade, blood and a blue apron. This fly has continued to catch fish on stillwaters and rivers since the day of its invention over 150 years ago.

The Gold Butcher has a gold tinsel body instead of silver.

Hook 10-14 (larger sizes for seatrout).
Thread Black.
Tail Red ibis substitute.
Body Flat silver tinsel.
Rib Oval silver tinsel.
Hackle Black.
Wing Blue mallard quill feather.

Bloody Butcher

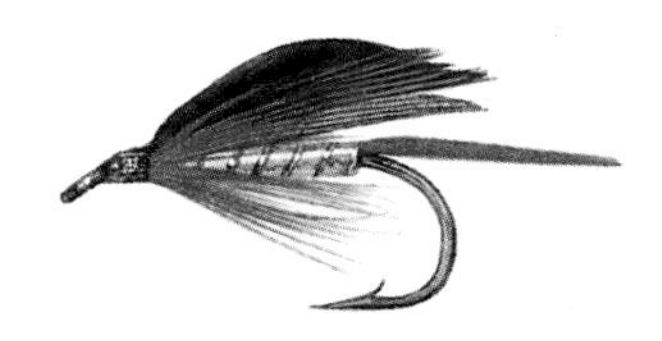

A variant of the previous fly, with a bright red hackle added. This is also a favoured seatrout fly.

Hook 10-14 (larger for seatrout).
Thread Black.
Tail Red ibis substitute.
Body Flat silver tinsel.
Rib Oval silver rib.
Hackle Scarlet.
Wing Blue mallard quill feather.

Zulu

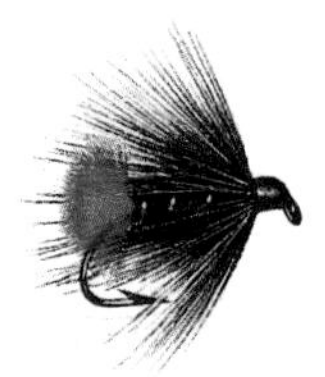

This again is a very old pattern, used for both lake and seatrout. In larger versions it is used as a dapping fly.

The alternative Blue Zulu dressing is given in the seatrout section.

Hook 8-14.
Thread Black.
Tail Red wool.
Body Black silk.
Rib Silver oval.
Hackle Palmered black cock.
Wing None.

Soldier Palmer

A top fly on such waters as Grafham, the Soldier Palmer goes from strength to strength. Usually fished as the top dropper on a team of three traditional flies, this fly accounts for many fish. It is possible that this pattern is the original artificial fly. Some anglers call for a red tail on this fly, but this is optional.

Hook 8-14.
Thread Black.
Tail Red wool (optional).
Body Red wool.
Rib Gold or silver wire.
Hackle Palmered medium red cock.
Wing None.

Mallard and Claret

This is probably the most effective of the Mallard series. It is a traditional pattern still used by today's anglers on stillwaters. Other flies in the series are Mallard and Yellow, Mallard and Blue, Mallard and Green, Mallard and Black, Mallard and Silver, Mallard and Orange, and so on. All that alters is the body and hackle colour.

Hook 8-14.
Thread Black.
Tail Golden pheasant tippets.
Body Claret seal's fur, or wool.
Rib Gold oval tinsel.
Hackle Black or claret.
Wing Bronze mallard shoulder.

Invicta

One of the most popular fly patterns ever conceived and still a firm favourite with both stillwater anglers and seatrout fishermen. The fly was created by James Ogden of Cheltenham, who originally came from Derbyshire. He is thought by some to be the father of the dry fly as we know it today. The Invicta is a superb pattern during a sedge fly rise, perhaps imitating a hatching caddis or a returning egg-laying female of a species that descends beneath the water surface to oviposit. The closely allied pattern, the Silver Invicta, is given later. Another modern version, the Red-tailed Invicta, has a red attractor tail.

Hook 8-14.
Thread Black or yellow.
Tail Golden pheasant crest.
Body Yellow seal's fur.
Rib Gold wire or oval.
Hackle Palmered light red cock with blue jay at the throat.
Wing Hen pheasant centre tail, or wing, which is a little easier to use.

Black Pennell

This is one of a famed series of flies devised by the Victorian author, Cholmondeley Pennell, for both trout and seatrout. The black version is extremely effective during a rise of chironomid midges. In larger sizes, it is considered by many to be a useful seatrout fly. The other colour variation still in use today is the Claret Pennell.

Hook 8-14.
Thread Black.
Tail Golden pheasant tippets.
Body Black silk.
Rib Silver oval tinsel, or flat if preferred.
Hackle Black cock or hen.
Wing None.

Coachman

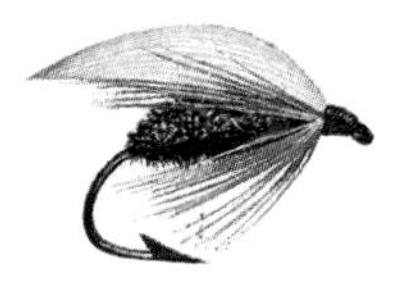

This is a particular favourite of mine. With its bright white wing, it has been particularly useful at dusk. I always fish it as a wet fly, although some anglers favour a dry version. The fly was created in the nineteenth century, supposedly by a coachman to the reigning monarch – but that's as maybe. It is thought by many to be an impression of a moth of sorts. A variant known as the Leadwing Coachman sports a more sombre grey wing.

Hook 8-14.
Thread Black.
Tail None.
Body Peacock herl.
Rib None.
Hackle Natural red cock or hen.
Wing White goose.

Cowdung

No visitor to the countryside, be he angler or not, can have failed to notice the hairy olive-coloured flies that crawl over and buzz around freshly dropped cowpats. These terrestrial nasties are sometimes blown onto the water, where the trout show no aversion to their previous habitat. The cowdung has been copied since the dawn of fly dressing, and is still used by anglers of the rough streams.

Hook 12-14.
Thread Black.
Tail None.
Body Peacock herl.
Rib None.
Hackle Light natural red.
Wing Light brown hen quill.

Royal Coachman

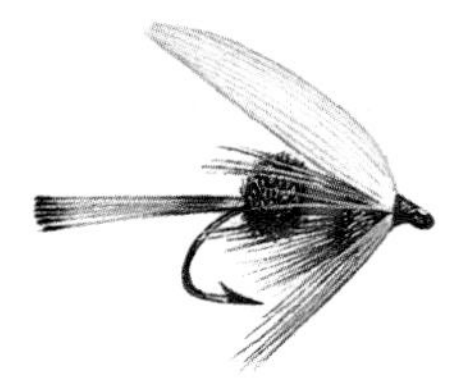

The humble Coachman fly was taken to the United States of America, where its livery was livened up with a red waistcoat. This fly is a standard wet pattern in the Americas. It has made the return journey to these shores, where it can prove quite useful as an attractor pattern in coloured water on both rivers and lakes.

Hook 8-14.
Thread Black.
Tail Golden pheasant tippet.
Body Bronze peacock herl with red silk centre.
Rib None.
Hackle Natural red cock or hen.
Wing White goose.

Grouse and Green

The Grouse series of flies is used for both trout and seatrout. All are furnished with a wing from grouse's tail. The Green is probably the most widely used, but the Grouse and Claret is often used as a substitute, or as an alternative to the Mallard and Claret. Others in the series include Grouse and Yellow, Grouse and Silver, Grouse and Gold, Grouse and Blue. Only the body and hackle colours are changed.

Hook 8-12.
Thread Black or green.
Tail Golden pheasant tippet.
Body Green seal's fur.
Rib Oval gold tinsel.
Hackle Natural red hen or green.
Wing Grouse tail.

Dunkeld

A few years ago I was a guest at an international fly-fishing match between England, Scotland, Ireland and Wales. Before the match, I wandered between the teams huddled together in secret conclave. All were secretly settling on flies to be used that day, and all had chosen the Dunkeld as one of their team of three flies, unbeknown to each other. This fly of Scottish origin is a firm favourite of the reservoir angler. The original fly called for cheeks of jungle cock, but nowadays they are usually omitted.

Hook 8-12.
Thread Black or orange.
Tail Golden pheasant crest.
Body Flat gold tinsel.
Rib Oval gold rib.
Hackle Palmered orange cock hackle.
Wing Bronze mallard, with jungle cock cheeks optional.

Green Peter

This fly is fished wet or dry, and represents a sedge. The origin of this pattern is firmly in the Emerald Isle, where it is used to great effect on the famed lakes of that country. Like the Bibio, it is a popular pattern amongst reservoir anglers.

Hook 8-12.
Thread Black or olive.
Tail None.
Body Light green seal's fur (pea green).
Rib Fine oval gold.
Hackle Palmered ginger cock, two ginger at the head.
Wing Hen pheasant, or light brown speckled hen.

Bibio

This fly is Irish in origin, but in recent years it has received wide attention throughout the British Isles. The Bibio is a group of true flies which includes the Black Gnat and the Hawthorn. Another version has a black/red/black/red body. This fly is best fished as a bob fly in the surface film.

Hook 10-14.
Thread Black.
Tail None.
Body Black seal's fur with a claret seal's fur centre.
Rib Silver wire or oval.
Hackle Palmered black cock.
Wing None.

Alexandra

Truly an attractor pattern, this was named after Queen Alexandra when she was a Princess. It was originally called the Lady of the Lake, evoking the Arthurian legends. In its early days it was considered too deadly and was banned on some waters. Today, though used by some die-hard lake fishermen, it is as a seatrout fly that the Alexandra comes into its own.

Hook 8-12.
Thread Black.
Tail Red ibis substitute and two or three peacock sword tail fibres.
Body Flat silver tinsel.
Rib Oval silver tinsel.
Hackle Black cock or hen.
Wing Green peacock sword flanked with strips of red ibis substitute (scarlet swan).

March Brown

The natural March brown (*Rhithrogena haarupi*) is extremely local in its incidence, occurring on very few rivers, and yet the artificial fly is fished in all corners of the British Isles. Most areas have their own individual patterns, the Red-legged March Brown, the Claret March Brown, even a Purple March Brown, to name but a few of the varieties to be found. The fly is an insect of the wild, flowing rocky rivers, not of the sedate chalkstreams, although it is found in Normandy. This is the reason why scholars believe that the *Treatyse* originated from that area, for a March Brown Imitation is given as the first fly in the list of twelve. The flashier Silver March Brown is given in the seatrout section.

Hook 8-14.
Thread Black or brown.
Tail Partridge tail fibres.
Body Hare's ear and body fur mixed.
Rib Gold wire.
Hackle Brown partridge.
Wing Hen pheasant wing slips.

Black Spider

This is not an imitation of a spider as such, but the word 'spider' indicates the style of dressing. A sparse mobile hackle that kicks and pulsates in the water is the hallmark of the spider-dressed flies. The Black Spider is a must for all rough stream fishing anywhere in the world.

Hook 12-18.
Thread Black.
Tail None (though some patterns do give a tail).
Body Black silk.
Rib None.
Hackle Black hen.
Wing None.

Peter Ross

A lake and seatrout fly of proven worth evolved by one Peter Ross of Perthshire in the late nineteenth century. He altered the already established Teal and Red fly and turned a good pattern into a more killing one, so that today, as far as the Teal series of flies is concerned, without doubt the Peter Ross heads the list.

Hook 8-12.
Thread Black.
Tail Golden pheasant tail fibres.
Body Two-thirds silver tinsel; one-third red seal's fur.
Rib Oval silver tinsel.
Hackle Black.
Wing Barred teal flank.

York's Special

This fly, known as the Yorkie in North Wales fishing circles, is a popular pattern for the lakes of the Principality. It was created by a gentleman by the name of York who travelled the area selling flies and other items of tackle to the local stores. The York's Special is a good fly to use when the heather fly (sometimes called the Bloody Doctor), which looks like a hawthorn with red legs, falls on the water.

Hook 10-14.
Thread Black.
Tail Red-dyed swan or goose.
Body Peacock herl.
Rib None.
Hackle Coch-y-Bonddu.
Wing None.

Gosling

An Irish pattern for the great limestone lakes of that country. This fly is an imitation of the mayfly and is fished wet or just in the surface film. This particular Gosling was created by Michael Rogan of Ballyshannon in the family fly-dressing business of that name. Rogans were famed for their exquisite salmon flies, which were all tied without the aid of a vice. In the mid-nineteenth century, Francis Francis described Rogan's fly as a 'piece of jewelery'.

Hook 8-10.
Thread Yellow.
Tail Cock pheasant tail fibres.
Body Golden olive seal's fur.
Rib Gold wire.
Hackle Orange cock with grey speckled mallard in front.
Wing None.

Woodcock and Yellow

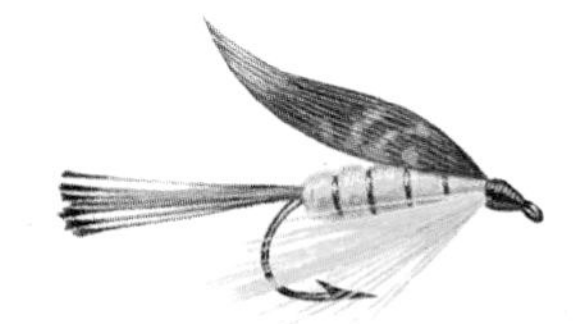

Probably the most popular of the Woodcock series of trout flies, the Woodcock and Yellow is used in small sizes as a river pattern, slightly larger for lake use, and in the largest sizes it is favoured as a seatrout pattern.

Hook 8-12.
Thread Yellow.
Tail Golden pheasant tippet.
Body Yellow seal's fur.
Rib Gold oval tinsel.
Hackle Light ginger or yellow.
Wing Woodcock wing slips.

Kell's Blue

This is a fly devised for the River Usk by the late Eddy Kelly. It is similar in many respects to the Usk Nailer, and even to the dry Kite's Imperial. On its day it has proved a very killing pattern.

Hook 12-14.
Thread Purple.
Tail Blue dun cock hackle fibres.
Body Heron herl.
Rib Silver oval.
Hackle Blue dun hen.
Wing None.

Poacher

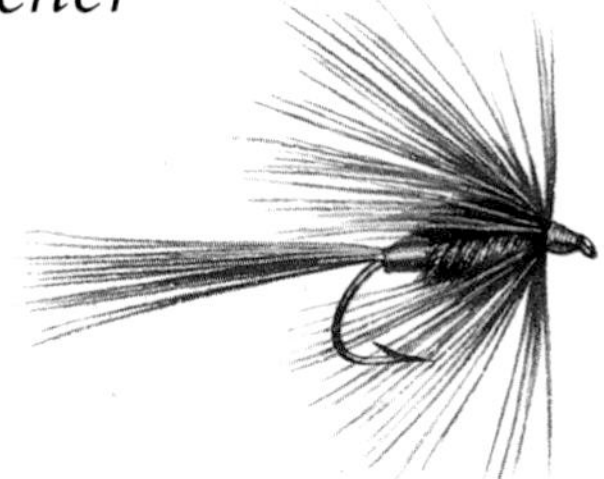

Originally from the Loch Lomond area of Scotland, this pattern has won wide acclaim as a bob fly on a team of three wet flies. Similar to the Coch-y-Bonddu, there are a number of variants in the dressing. Some have a red wool tail and a body of equal halves of orange seal's fur and peacock herl. The fly depicted here has a tip of orange floss and a tail of golden pheasant red body feather.

Hook 8-14.
Thread Black or brown.
Tail Golden pheasant red body feather fibres.
Body Bronze peacock herl with a tip of hot orange silk.
Rib None.
Hackle Dark furnace (natural red with black centre).
Wing None.

Diawl Bach

As its name indicates, this is a Welsh fly. Diawl Bach translated means Little Devil. Though Welsh by birth, this pattern has achieved its greatest success on such waters as Chew Valley and Blagdon reservoirs.

Hook 12-14.
Thread Black or brown.
Tail Brown cock hackle fibres.
Body Peacock herl.
Rib None.
Hackle Dark brown hen.
Wing None.

Black and Peacock Spider

Another simple, hackled wet fly that goes back to the days of early fly tying. It was brought to light and popularized by Tom Ivens, one of the pioneers of modern reservoir fishing. Some believe that the fly represents a species of aquatic snail, and they may well be right. I am inclined to believe that the trout take this fly because it looks edible, and can represent a wide number of small creatures, both aquatic and terrestrial. I am never without this pattern when I fish the wilder lakes or reservoirs of the UK.

Hook 8-14.
Thread Black.
Tail None.
Body Peacock herl.
Rib None.
Hackle Black hen.
Wing None.

Blae and Black

A traditional lake pattern originating in Ireland, but now used all over the British Isles. It is particularly effective during a rise of chironomid midge. In some places this fly is known as the Duck Fly. It can also be a useful fly for seatrout.

Hook 8-12.
Thread Black.
Tail Golden pheasant tippets (alternatively, black cock hackle fibres).
Body Black floss.
Rib Oval silver tinsel.
Hackle Black cock or hen.
Wing Grey duck.

Cinnamon and Gold

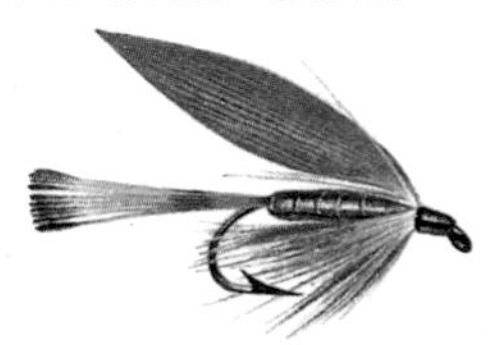

A favoured loch pattern from Scotland, this fly has also seen sterling service in the armoury of the seatrout angler. There are a number of variants in the dressing. Some patterns use golden pheasant tippet for the tail; others ginger hackle fibres. Yet another variety has a red ibis tail.

Hook 8-12.
Thread Brown.
Tail Golden pheasant tippets, or light ginger hackle fibres.
Body Flat gold tinsel.
Rib Oval gold.
Hackle Ginger.
Wing Cinnamon hen quill, or turkey in larger sizes.

Kingfisher Butcher

The fanciest of the Butcher range of lake patterns, the touch of orange makes this one very attractive to the modern reservoir rainbow trout. There is a slight difference of opinion regarding the wing. Some dressings call for a blae (grey) wing. Others, in keeping with the Butcher profile, give the normal blue mallard feather wing.

Hook 8-12.
Thread Black.
Tail Blue kingfisher feather fibres.
Body Flat gold tinsel.
Rib Oval gold tinsel.
Hackle Hot orange cock or hen.
Wing Blue mallard slips (sometimes a grey wing is given).

Watson's Fancy

Another fancy pattern favoured in both Scotland and Ireland for lake fishing. Like many of the traditional loch patterns, it is also a useful seatrout fly. The Jungle cock cheeks are now, of course, optional due to the protection of the species. However, in the UK birds are bred and reared solely for their feathers, which are plucked without harm to the birds, so once more many of the 'Jungle cock' flies can be seen in their full glory.

Hook 8-10.
Thread *Black.*
Tail Golden pheasant crest.
Body Two halves red seal's fur followed by black seal's fur.
Rib Oval silver tinsel.
Hackle Black hen.
Wing Black-dyed goose or natural crow, cheek Jungle cock.

Grizzly King

A standard North American pattern which is also used in Canada. This fly cannot be considered an original American fly, for it was created by John Wilson, Professor of Philosophy at Edinburgh University, Scotland. Since its inception, the pattern has been tied up successfully as a streamer fly.

Hook 8-12.
Thread Black or green.
Tail Red-dyed swan or goose.
Body Green seal's fur.
Rib Oval gold.
Hackle Grizzle hen.
Wing Grey speckled mallard.

Rio Grande King

Another royal fly from the USA, and like the Grizzly King it too has received the long-shank treatment, and has proved effective as a streamer pattern.

Hook 8-12.
Thread Black.
Tail Yellow cock hackle fibres.
Body Black chenille tipped with flat gold tinsel.
Rib None.
Hackle Brown hen.
Wing White duck or goose.

Picket Pin

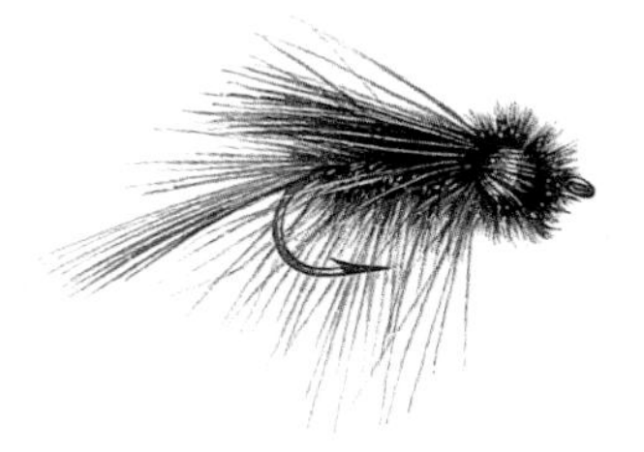

Though this American pattern is classed as a wet fly, fished on a sunk line slowly along the bottom, it has taken a number of good trout for me. It is possible that the trout mistook the fly for a dragonfly larva. It is best to weight the pattern.

Hook Long shank 8-12.
Thread Black.
Tail Brown cock hackle fibres.
Body Bronze peacock herl.
Rib None.
Hackle Palmered with brown hackle.
Wing Grey squirrel. A head of peacock herl is tied in after the wing.

Dr Sprately

I first came across the North American pattern in Wales, of all places, where a number of my friends were taking good bags with this fly. Sometimes known as Doc Sprately, or just Sprately, this fly was named after Dr Donald A. Sprately from Mount Vernon, Washington State, and was created by Dick Prankard around 1949. This fly was one of the most popular patterns used in British Columbia.

Hook Long shank 8-10.
Thread Black.
Tail Grey hackle fibres.
Body Black wool.
Rib Flat silver tinsel.
Hackle Grizzle.
Wing Pheasant tail fibres. A peacock herl head is tied in after the wing.

Burlap

This simple, hackled pattern is used throughout the Pacific North West for a wide variety of game fish, in particular the migratory steelhead.

Hook Long shank 8-2.
Thread Light brown.
Tail Bunch of brown bucktail.
Body Natural burlap (sack hessian) well picked out.
Rib None.
Hackle Long-fibred, soft grey grizzle cock or hen.
Wing None.

Montreal

A standard American classic fly for all species of trout, and also black bass.

Hook 8-12.
Thread Black.
Tail Dyed red hackle fibres.
Body Claret floss.
Rib Flat gold tinsel.
Hackle Claret cock or hen.
Wing Brown mottled hen, or turkey in larger sizes.

Parmachene Belle

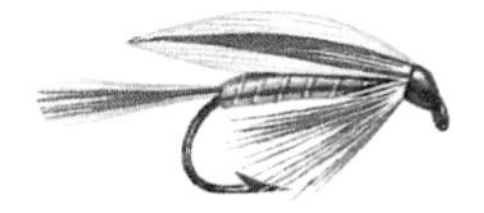

Probably one of the best-known American wet-fly patterns. With its striking red and white garb, it is easily recognized by most anglers. It is an attractor fly pure and simple, and has a number of devotees among the British still-water brigade.

Hook 8-12.
Thread Black or yellow.
Tail Red and white cock hackle fibres.
Body Yellow floss.
Rib Flat gold tinsel.
Hackle Mixed red and white cock.
Wing White goose with centre strip of red-dyed goose.

Orange Woodcock

This fly hails from New Zealand, where it fills the role of such British flies as the Partridge and Orange.

Hook 10-16.
Thread Orange.
Tail None.
Body Orange wool.
Rib Fine silver tinsel.
Hackle Woodcock hackle.
Wing None.

Temuka

This New Zealand wet fly is probably taken by the trout for some form of beetle. It has many similarities with the famous British fly, the John Storey.

Hook 8-12.
Thread Black.
Tail None.
Body Peacock herl.
Rib None.
Hackle Claret.
Wing Grey speckled mallard flank.

Carter's Pink Lady

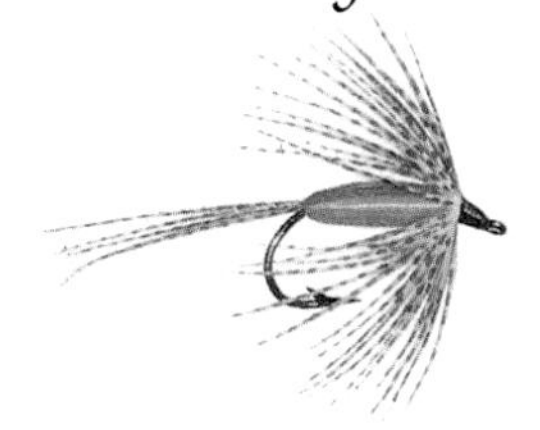

I brought back a number of specimens of this fly from South Africa and, never one to let anything go to waste, have used them on my local waters, where they have behaved no worse than any other fly. I have since tied a few specimens with a fluorescent pink floss for seatrout.

Hook 8-10.
Thread Black.
Tail Speckled hen fibres.
Body Pink wool.
Rib None.
Hackle Speckled hen hackle.
Wing None.

Kenya Bug

Here again is a pattern that could well fall into the nymph category. As its name suggests, it is used right up into Kenya.

Hook 8-12.
Thread Black.
Tail Black cock hackle fibres (alternatively, a blue guinea fowl).
Body Black wool.
Rib Silver tinsel.
Hackle Long-fibred black hen.
Wing None.

Taddy

An excellent South African fly that imitates the young frog. Trout in the dams quite often selectively feed on these creatures. It is then that the Taddy comes into its own.

Hook 8-12.
Thread Black.
Tail and body Both made from the same bunch of black squirrel hair.
Rib None.
Hackle None.
Wing None.

Smart's Yellow Tail

This is a fancy fly. Because of its palmered hackles it can work as a good bob fly, although it probably was not fished thus in South Africa. A very attractive pattern.

Hook 8-10.
Thread Black or yellow.
Tail Golden pheasant crest.
Body Yellow seal's fur.
Rib Gold wire.
Hackle Palmered honey cock. Apple green followed by grey partridge, and finally a white cock or hen hackle.
Wing None.

Pallaretta

A traditional fly from Spain that is fished both sides of the Pyrenees. The feathers for most Spanish flies come from the fabled coqs de Leon, whose superb spade feathers have a glass-like finish. The colour for the Pallaretta hackle is called Indio Acerado (steel grey). In many of the traditional Spanish wet flies the hackle does not go completely around the hook, only above and around the sides.

Hook 12.
Thread Black or brown.
Tail None.
Body Yellow silk varnished.
Rib Black silk.
Hackle Dark blue dun.
Wing None.

Oliva

As its Spanish name indicates, this fly imitates one of the many olive Ephemerids. This fly breaks with the Spanish tying tradition in having a hackle all around the shank.

Hook 12-14.
Thread Primrose.
Tail None.
Body Medium olive fur or synthetic dubbing.
Rib None.
Hackle Soft grizzle hen.
Wing None.

Mave

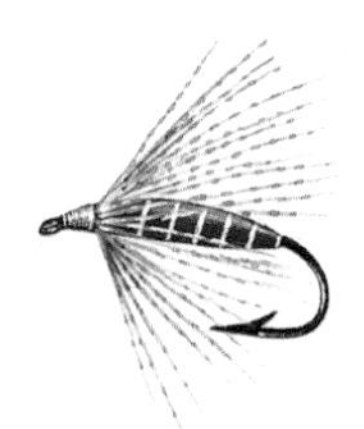

Another Spanish fly tied in the traditional Spanish wet-fly style. This time the feather used for the hackle is called Flor de Escoba (of a colour of the Flower of the Mountain Broom). All mottled feathers are prefixed by the word Prado, and plain hackles are prefixed Indio. If a white hackle is used (Indio Palometa) then another traditional fly, the Albernios Pena, can be tied.

The Mave could be classed as the Snipe and Purple of Spain.

Hook 12-14.
Thread Pale yellow (primrose).
Tail None.
Body Purple.
Rib Primrose.
Hackle Mottled spade hackle fibres (Flor de Escoba).
Wing None.

The Valasesiana

This type of fly comes from the Sesia valley in the Piedmont region of Northern Italy, bordering on Switzerland. These patterns have been in use on the River Sesia for over 200 years, the first reference to them appearing in church chronicles around 1760.

They were originally fished using hazel or ash poles, about 3¾-4½ yd (3.5-4 m) long, with a line of braided male horsehair, a gut leader, and four flies. The flies were used for both trout and grayling: sizes 10-16 for trout, and 16-18 for grayling. The bodies were made from various shades of cotton or silk, and they had no tails. The hackles came from a wide variety of young birds, including starling, blackbird, thrush, owls and domestic fowl no more than six months old. The flies were finished behind the hackle, giving it a forward tilt. This provided the fly with plenty of movement in the fast-running rivers of that region. This style of fly and fishing will probably die out within the next ten years.

These Italian flies are tied on blind (eyeless) hooks. To tie on eyed hooks, an ideal model is the Partridge 'Grub' hook.

Hook 16-18 for grayling; 10-16 for trout.
Thread Red.
Tail None.
Body Red silk.
Rib None.
Hackle Grey/blue dun.
Wing None.

Hook 16-18 for grayling; 10-16 for trout.
Thread Olive.
Tail None.
Body Olive.
Rib None.
Hackle Starling neck hackle.
Wing None.

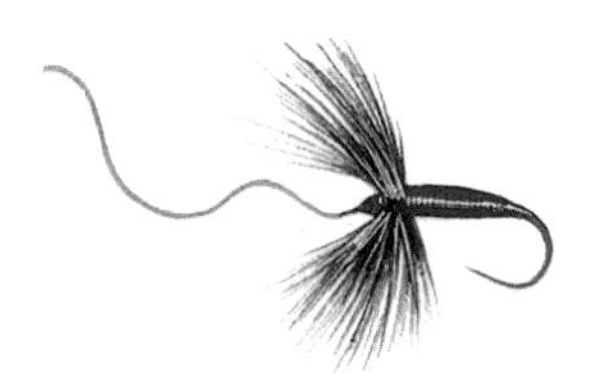

Hook 16-18 for grayling; 10-16 for trout.
Thread Purple.
Tail None.
Body Purple.
Rib None.
Hackle Black hen.
Wing None.

Hook 16-18 for grayling; 10-16 for trout.
Thread Black.
Tail None.
Body Black.
Rib None.
Hackle Black hen.
Wing None.

The Ossolina

These flies hail from the Ossola valley, to the west of the Sesia valley, also in the Piedmont region of Italy. They were mainly fished in the River Toce and other small streams of that area. They originated in the town of Domobossala, the largest town in the area, and are unique in the fly-tying world as they are partially made by machine. It appears that there are only two old men left in the area tying flies in this traditional way.

In the old days the area was renowned for its needles, and it was with the rejects from the needle factories that the flies were made. Needles of the correct size are placed in a rotating vice, which is activated by a foot pedal. Any tails and hackles are tied on first, and then the body is wound on by foot power. The straight needles are taken out of the vice and bent around a special former to produce the hook. Thus these flies are the only ones that are in the world tied before the actual hook is formed.

Like the Valasesiana flies they are used for both trout and grayling. Both types of fly are fished wet or in the surface film, usually downstream.

Ossolina Nymph Style

Hook 16-18.
Thread Brown.
Tail Cock pheasant tail fibres.
Body Cock pheasant tail fibres.
Rib Gold wire.
Hackle A few wisps of grey feather emerging from the head of the fly.
Wing None.

Ossolina Emerger or Spent 2

Hook 14-16.
Thread Black.
Tail None.
Body Naples yellow (dirty yellow).
Rib None.
Hackle Sparse long-fibred pale ginger.
Wing None.

Ossolina Emerger or Spent 1

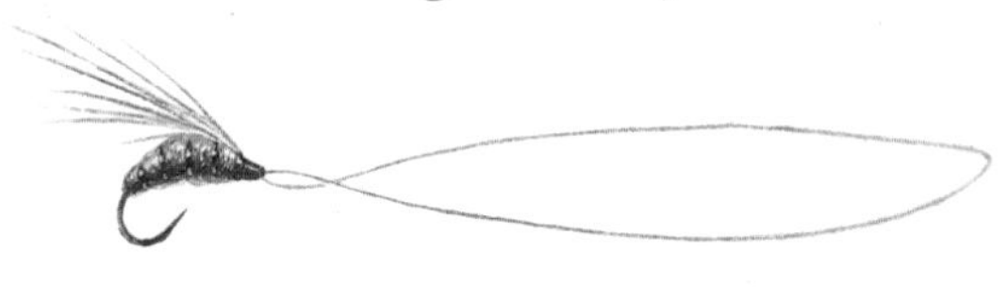

Hook 14-16.
Thread Red or black.
Tail None.
Body Tan-coloured silk or cotton.
Rib Thin flat silver.
Hackle A few wisps of light ginger cock hackle.
Wing None.

Ossolina Dun or Spinner 1

Hook 16.
Thread Black.
Tail None.
Body Brown silk or cotton.
Rib Ribbed with fine silk or fine silver wire.
Hackle Soft ginger hen hackle.
Wing None.

Ossolina Dun or Spinner 2

Hook 16.
Thread Black.
Tail None.
Body Brown silk with yellow tip.
Rib None.
Hackle Sparse black hen or other black feather.
Wing None.

Bartellini Spiders

The following three patterns were created by Walter Bartellini of Turin. They are just three taken from a series of about forty spider-dressed patterns. They are used for both trout and grayling on such rivers as the Orco, Po, and Stura. They differ from the other Italian patterns given in the use of more modern materials, such as fluorescent silks. There is an interesting similarity between the Bartellini series of flies and flies used in the Spanish Pyrenees. In most cases, the emphasis is on the colour of the heads of these flies. Quite often the heads are of a totally different colour to the body and to the overall shade emphasis of the fly. This colour feature is not apparent in the heads of other European or American spider flies.

Red/White

Hook Grub type 16-20.
Thread Red.
Tail None.
Body Red fluorescent silk.
Rib None.
Hackle White hen.
Wing None.
Head Fluorescent green.

Brown/Peacock

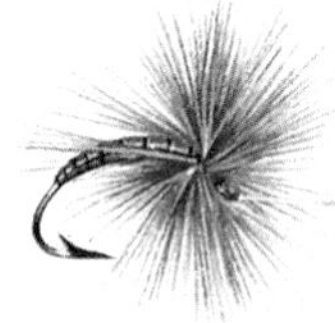

Hook Grub type 16-20.
Thread Black.
Tail None.
Body Stripped peacock quill.
Rib None.
Hackle Brown hen.
Wing None.

Blue/Grey

Hook Grub type 16-20.
Thread Red.
Tail None.
Body Dark blue silk.
Rib Fine flat silver tinsel.
Hackle Smoke-grey hen.
Wing None.
Head Red.

NYMPHS AND PUPAE

At one time the term 'nymph' was used to describe only the larval stage of the Ephemeroptera (the mayflies) and the larvae of the Odanata (damsel and dragonflies). The nymphs of the latter were sometimes also referred to as naiads.

The larval stage of other aquatic insects had their own generic tems: Plecoptera (stoneflies) were called creepers, and Tricoptera (sedge flies) were called caddis worms. The infant stages of the water-boatmen and the corixae were referred to as instars.

The term 'nymph' is now far broader in meaning, for it has come to describe and include those flies that have been specifically designed to imitate any creature that lives below the water. Larval and pupal stages, and even adults of some creatures, are embraced by the term 'nymph', as are freshwater shrimps and the lowly water-louse, water-boatmen and corixae, the voracious larvae of the water beetles, and the wide range of larvae and pupae of many of the aquatic true flies (Diptera). Fishing such artificials is termed 'nymph fishing'.

It is difficult to draw the line between fishing the nymph and fishing the wet fly, as in some cases it is hard to differentiate between the two. Some wet flies are nymphs in concept, though not in design. Other wet flies can look very much like nymphs, and certainly some nymphs could well be described as wet flies. One man's nymph is another man's wet fly, and vice versa.

Nymph fishing started in the latter half of the nineteenth century with the enquiring mind of G. E. M. Skues. Skues flaunted his new nymphs in the face of the dry-fly dogma of his day, earning the wrath of many because he caught more fish with them.

Most anglers confined their sport to casting a fly to a rising fish. When rises were not apparent on the surface, they did not fish; such was their angling Puritanism. Skues, on the other hand, contended that the trout fed avidly below the surface on the hatching nymph and were sometimes content to do this rather than rise to the floating adult fly. Upstream wet-fly fishing as practised by Northern anglers bore witness to this. All Skues did was to adapt these wet-fly tactics to the chalkstreams, and developed his nymph flies to suit the water. The story goes that he was fishing the dry fly, a badly tied one as it happens, which sunk straight away. As soon as it went below the surface it was taken by a trout. The trout continued to take the sunk fly, ignoring any floating pattern. The importance of this observation was not apparent to Skues at the time, but later it dawned on him, and he mentions this in his book, *Minor Tactics of the Chalkstream*.

Nymph fishing is widely practised by most fly-fishermen; on British chalkstreams in recent years the names of Frank Sawyer and Oliver Kite have become synonymous with fishing the nymph. In the USA notable anglers such as James Leisenring pioneered the use of the nymph on the rivers of that country. Today the importance of the nymph is well appreciated by most anglers and the great influx of stillwater anglers to the fishing scene has possibly made nymph fishing the prime method of taking fish.

The nymph patterns of this section are for both river and stillwater, and show a great variety in size, and in the materials used to create them. They cover most aquatic creatures, and even include some nymphs from the realms of fancy, which imitate nothing in particular and yet are proven fish-takers.

Bloodworm

This pattern is designed to imitate the larval stage in the life cycle of the chironomid midge. This creature can vary in colour; some are olive green, some yellow, with a high proportion blood red, hence the name. Because the creature lives buried deep in the mud, where the dissolved oxygen level is low, nature has provided the larva with haemoglobin so that it can retain oxygen in its system. From time to time, it rises to a better-oxygenated level in order to replenish its supply. During these forays the larva receives the attention of the trout. The red marabou is tied not just as a tail, but also as an extension of the body, in order to give a wriggling motion to the fly.

Hook Long shank 12-14.
Thread Red.
Tail Tuft of red marabou.
Body Red floss silk tied with distinct undulations.
Rib Fluorescent red floss.
Hackle None.
Wing None.
Head Bronze peacock herl.

Buzzer Pupa

This is the most important stage in the life cycle of the midge, as far as the trout is concerned. There are many colour variations of this fly, including olive, red, orange, bright green, and brown. The one depicted here is the black. The chironomid midge is perhaps the major item in the stillwater trout's diet.

Hook 10-16.
Thread Black.
Tail Tuft of white fluo floss.
Body Black floss silk.
Rib Silver wire.
Hackle None.
Wing None.
Thorax Bronze peacock herl.
Breathing filaments Tuft of white fluo floss or white feather fibre.

Buoyant Buzzer (Suspender Buzzer)

Just prior to hatching, the pupa hangs motionless in the surface film, before adopting a horizontal position for the hatching. This style of fly was devised originally by the American fly-rodder and fly-tyer, Charles E. Brooks, and was further developed by John Goddard, the British angling entomologist.

Hook 10-16.
Thread Black.
Tail White DFM silk or wool.
Body Black seal's fur or silk.
Rib Silver wire.
Hackle None.
Wing cases Orange goose feather slips.
Thorax Peacock herl.
Head Bead of pastazote or polystyrene wrapped in nylon mesh.

Yellow Corixa

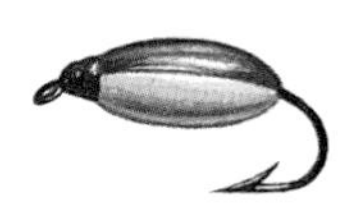

Many of the instars of the water bug corixa, and for that matter the water-boatman, are a totally different colour to the final adult. In some instances they are a definite yellow colour. It is this infant stage that this fly of Richard Walker's represents.

Hook 12-14.
Thread Yellow.
Tail None.
Body Yellow floss weighted beneath with lead strip.
Rib None.
Hackle None.
Wing case Olive green feather tied across the back, then the whole fly is varnished.

Silver Corixa

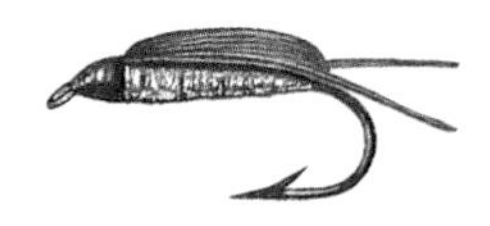

This small water bug swims swiftly up to the surface in order to trap air on the underside of its abdomen. This bubble of air is then taken back down under the water. The air appears like a coating of silver beneath the insect.

Hook 12-14.
Thread Black.
Tail None.
Body Flat silver tinsel.
Rib Oval silver tinsel.
Hackle None.
Wing case (back) Cook pheasant tail fibres.
Paddles Two cock pheasant tail fibres.

Chomper

This is a broad-spectrum, bug-type nymph that can be tied in a wide variety of colour combinations. The Chomper can imitate corixae, water beetles, and other creatures such as shrimps. It is a useful pattern to have for stillwater fishing.

Hook 12-14.
Thread Black.
Tail None.
Body Ostrich herl of appropriate colour.
Rib None.
Hackle None.
Back Raffene (Swiss straw) of appropriate colour.

Damsel Nymph

Yet another creation of the late Richard Walker. There are many damsel fly imitations and this one is a fair representation, and a proven killer. The damsel fly is a very good fly to use during the summer months on any of the smaller stillwater fisheries. As soon as you see the slim blue damsels on the wing, you will know it is time to fish the Damsel Nymph.

Hook Long shank 8.
Thread Olive or brown.
Tail Cock pheasant tail fibres dyed olive green.
Body Weighted first with lead strip. Mixed cobalt blue and orange lamb's wool. (Synthetic dubbing can be used.)
Rib Brown silk.
Hackle Grey partridge hackle dyed grass green.
Wing None.

Dragonfly Nymph

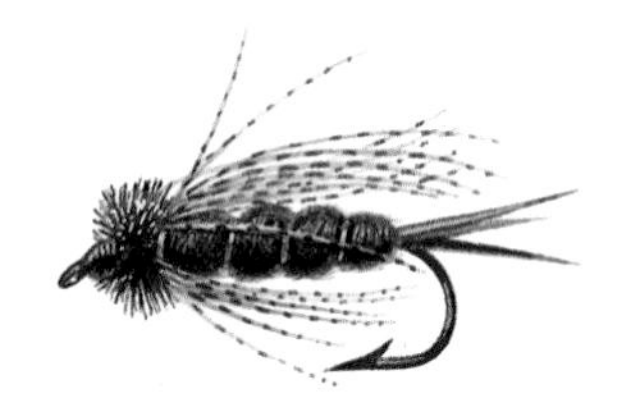

I tied this fly up a number of years ago, and it still continues to catch fish. The beauty of this fly is its simplicity of construction. It is fished in the same way as the Damsel Nymph: very slowly with occasional spurts.

Hook Long shank 8.
Thread Black or brown.
Tail Three short brown or olive goose biots.
Body Mixed dark brown/olive wool.
Rib Fluo green silk.
Hackle Brown partridge hackle.
Wing None.
Head Peacock herl.

Walker's Mayfly Nymph

This was one of the best patterns created by Richard Walker in the last few years before his untimely death. This fly is extremely popular on many of the clear stillwater fisheries, such as Avington, in Hampshire.

Hook Long shank 8.
Thread Brown.
Tail Four or five strands of cock pheasant tail fibre.
Body Underbody lead foil. Creamy yellow angora wool.
Rib Brown silk.
Legs The fine ends of the cock pheasant tail fibres used for wing case.
Wing case Cock pheasant tail fibres.
Thorax Same as body.

Alder Larva

Brian Clarke in his book, *Pursuit of Stillwater Trout*, described the Alder Larva as the Genghis Khan of stillwater. It is a voracious predator, but during the months of late March and April it feels the pupating urge. It then leaves the sanctuary of the detritus, and migrates to the bank of the river or lake in order to dig a small chamber in which to pupate. During this migration to the bankside it is preyed on by the trout. This particular pattern was devised by C. F. Walker.

Hook Long shank 10-12.
Thread Brown or black.
Tail Honey cock hackle tip.
Body Mixed brown and ginger seal's fur or substitute.
Rib Oval gold tinsel.
Hackle Honey hen hackle palmered; head hackle brown partridge.
Thorax Hare's ear.

Cove's Pheasant Tail

This fly was created by the renowned nymph-fisherman Arthur Cove, one of Britain's leading stillwater anglers, specifically for fishing from the banks of the large reservoirs. It is fished on a long leader, and with a slow retrieve. It can represent a wide number of sub-aquatic pupae.

Hook 8-14.
Thread Black or brown.
Tail None.
Body Cock pheasant tail fibres.
Rib Silver oval, gold or copper wire.
Hackle None.
Wing case Cock pheasant tail fibres.
Thorax Rabbit fur.

Pheasant Tail Variant

This is a long-shank version of the Cove's nymph. It can be tied with a variety of different-coloured thoraxes. The fly imitates a number of large aquatic larvae.

Hook Long shank 8-10.
Thread Black or brown.
Tail Cock pheasant tail fibres.
Body Cock pheasant tail fibres.
Rib Gold, silver or copper wire or fine oval.
Hackle None.
Wing case Cock pheasant tail.
Thorax Fluo green seal's fur, or orange, or yellow, and so on.

Sawyer's Pheasant Tail

One of the classic flies of the chalkstream, devised by the late Frank Sawyer, the world-renowned 'Keeper of the Stream'. He looked after his stretch of the River Avon in Hampshire, UK, for many years and became an acknowledged authority on the ways of the trout. His flies were created for the induced-take method of fishing.

Hook 12-16.
Thread Brown.
Tail Cock pheasant tail fibres.
Body Cock pheasant tail fibres wound with copper wire.
Rib None.
Hackle None.
Wing case Cock pheasant tail fibres.
Thorax A ball of copper wire covered with cock pheasant tail fibres.

Green Longhorn

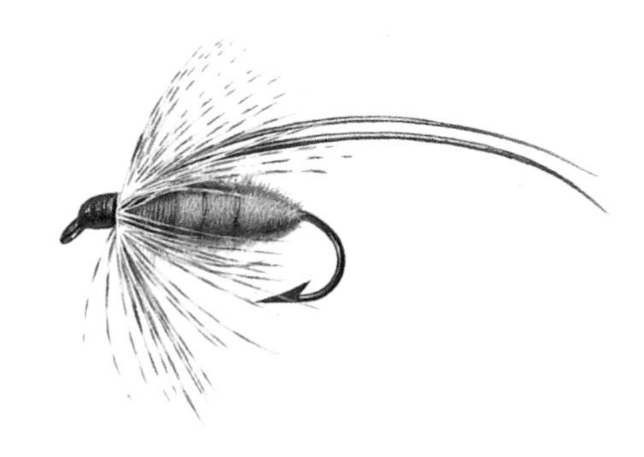

This fly is one of many sedge pupae patterns. The actual sedge/caddis pupa is a quiescent creature, lying in its shelter while changing into an adult in much the same way as a caterpillar rests in a chrysalis. The correct term for such flies is hatching sedge or caddis, rather than sedge pupa. There are a number of different body colours for this fly: green, orange/amber, cream and brown.

Hook 10-12.
Thread Yellow.
Tail None.
Body Sea-green dyed lamb's wool on rear two-thirds; one-third sepia lamb's wool.
Rib Fine gold oval over green part of body only.
Hackle Brown partridge.
Horns Two cock pheasant tail fibres.

Swannundaze Sedge Pupa

This is another version of the Longhorn. I have used a modern plastic strip called 'Swannundaze' introduced to fly-dressers by Frank Johnson of Lyndhurst, New Jersey, USA. In this fly, by using a fluorescent underbody of red silk, with the amber Swannundaze over it, one gets an almost blood-like effect in the abdomen of the fly. I first published this pattern in *Trout Fisherman*, January 1982.

Hook Shrimp hook size 8-10.
Thread Yellow.
Tail None.
Body Underbody fluo red silk; overbody clear amber Swannundaze.
Rib Strand of peacock herl between Swannundaze turns.
Hackle Brown partridge.
Wing case Cock pheasant tail fibres.
Thorax Brown seal's fur or substitute.
Horns Cock pheasant tail fibres.

Grey Goose

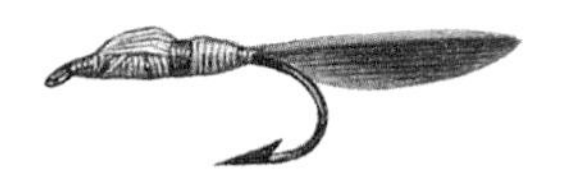

The second of Frank Sawyer's quartet of river nymphs. This pattern is used instead of the Pheasant Tail when a lighter pattern is required. Frank Sawyer did not choose the materials for his flies because they imitated natural nymph colours, but rather because the trout found them attractive.

Hook 12-14.
Thread Grey or white.
Tail Grey goose fibres.
Body Grey goose.
Rib None, but copper wire is wound with the goose fibres.
Hackle None.
Thorax Grey goose fibres.

Shrimper

The freshwater shrimp (*Gammarus pulex*) is an important item in the diet of the trout, in both rivers and stillwaters. There are many 'shrimp' flies. This one is sold commercially by a number of fly-dressing houses.

Hook 10-14.
Thread Brown or olive.
Tail None.
Body Olive seal's fur weighted with lead.
Rib Orange silk.
Hackle Palmered olive cock hackle.
Back Clear polythene.

Golden Shrimp

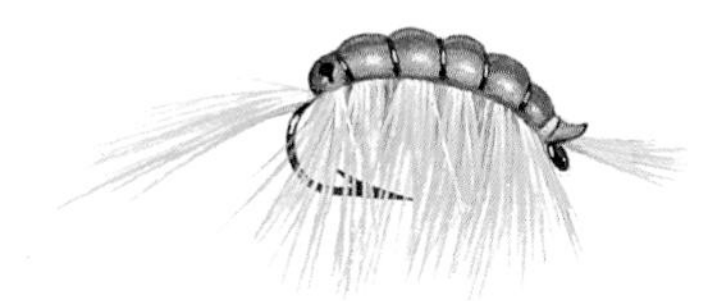

Freshwater shrimps come in a wide range of colours, such as grey and olive. Some have a distinct orange glow. And often during moulting, and certainly when dead, they are a yellow colour. Yellow seems to be a good colour to use when the water is muddy. The fish take it well in these conditions.

Hook Shrimp hook 8-10.
Thread Yellow.
Tail None.
Body Golden yellow seal's fur.
Rib Lead wire.
Hackle Golden yellow.
Back Yellow raffene or latex.
Feelers Yellow cock hackle fibres.

Colyer's Green Nymph

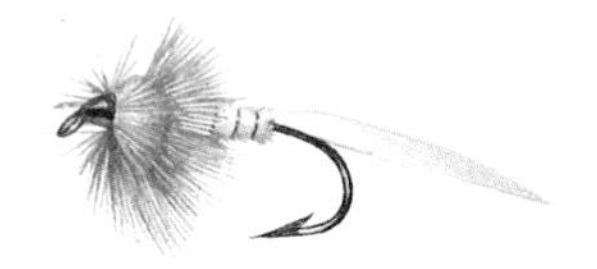

This broad-spectrum nymph is one of a series devised by Dave Collyer, the fly-dresser and author of *Fly Dressing 1* and *Fly Dressing 2*. Apart from the Green Nymph depicted here, there is a Black Nymph, Grey Nymph and Brown Nymph. All are constructed in the same way.

Hook 10-14.
Thread Olive.
Tail Olive goose or swan fibres.
Body Olive goose or swan.
Rib Oval gold tinsel.
Hackle None.
Wing case Olive goose fibres.
Thorax Olive ostrich herl.

Poly Shrimp

I based this fancy shrimp pattern on a more realistic fly created by Darrel Martin of Tacoma, USA. I used various colours of Fly Rite poly dubbing to create a range of fancy shrimps that owe very little to nature, yet have caught their fair share of fish. Colours tied: orange, black, green, grey, olive and pink.

Hook Shrimp hook 8-10.
Thread Green.
Tail Polypropylene.
Body Green seal's fur or poly dubbing.
Rib Fine gold.
Hackle Palmered green.
Back Green polypropylene.
Eyes Small metal bead chain or tiny beads.
Feelers Green cock hackle fibres.

Amber Nymph

This was one of the first flies devised to imitate a sedge pupa. The fly was created by Dr Bell for fishing Blagdon Water, one of the first man-made waters to offer top-quality fly-fishing for trout. There are two Amber Nymphs, the larger having a dark brown thorax and the smaller version an orange thorax. Both are popular patterns for the reservoir angler.

Hook 10-12.
Thread Black.
Tail None.
Body Amber seal's fur.
Rib None.
Hackle Sparse false hackle of ginger hen hackle fibres.
Wing case Grey/brown feather.
Thorax Brown seal's fur.

Barrie Welham Nymph

The original name for this nymph was the 'BW' (Brown Wool), but as this name coincided with the initials of the creator of the fly, the name was soon changed to the Barrie Welham. It is a useful fly when fish are taking chironomid pupae. The coloured tail provides an extra attraction.

Hook 10-14.
Thread Black.
Tail Mixed red and yellow fluo fibres, feather or hair clipped short.
Body Brown wool.
Rib Gold oval fine.
Hackle None, but a tuft of white feather at the top of the hook emerging from the head to act as breathing filaments.
Wing None.

Pond Olive Nymph

This simple nymph imitates the larval stage of the ephemerid, *Cloeon dipterum*. This tiny fly can hatch out in very large numbers on some British stillwaters.

Hook 14.
Thread Olive.
Tail Olive cock hackle fibres.
Body Dark olive seal's fur or synthetic substitute.
Rib Fine gold wire.
Hackle None.
Wing case Pale olive goose or swan fibre.
Thorax Dark olive seal or substitute.

Iron blue Nymph

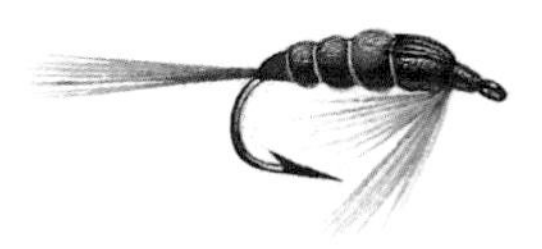

A river nymph designed to imitate the natural iron blue nymph stage, *Baetis pumilis* or *B. niger*. It is a fly that hatches out in very large numbers on many of the rivers in the UK.

Hook 14-16.
Thread Black or claret.
Tail Iron blue cock hackle fibres.
Body Mole fur.
Rib Fine silver wire.
Hackle None.
Wing case Black feather fibre.
Thorax Mole fur.

Sawyer's Swedish

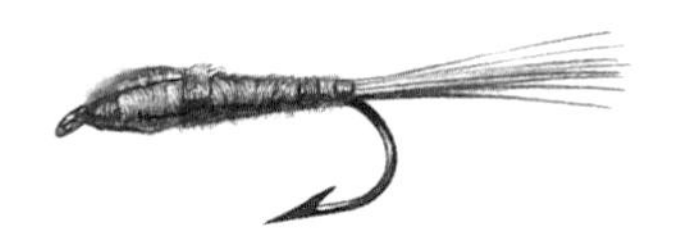

The third of Frank Sawyer's river nymphs, this pattern is darker than the Grey Goose Nymph. It is tied in exactly the same way as his Pheasant Tail Nymph.

Hook 12-14.
Thread Black.
Tail Dark goose fibres.
Body Dark goose fibres.
Rib None, but dark copper wire wound with body herl.
Hackle None.
Wing case Dark goose fibres.
Thorax Dark goose fibres.

Orange Nymph

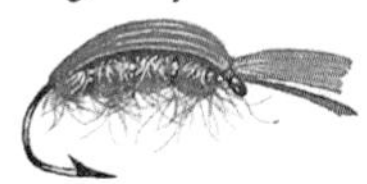

I devised this pattern to imitate, albeit very much larger, a form of Daphnia. Trout often become pre-occupied with feeding on these minute creatures, swallowing them down in their hundreds, and the Orange Nymph has proved successful on such occasions. Could it be that the trout are tempted because they see it as a veritable giant of a Daphnia? It is impossible to imitate the exact size of the Daphnia, they are far too small. Besides, if you could, your imitation would be lost in the veritable mass of real Daphnia.

Hook 14-16.
Thread Orange.
Tail None.
Body Orange seal's fur.
Rib Gold wire (optional).
Hackle None.
Wing case (back) Orange swan or goose feather tied across the back.
Feelers Two strands of goose left unclipped.

Grenadier Nymph

There are some who consider this to be a wet fly rather than a nymph pattern. It is another fly devised by Dr Bell, and it is still used very successfully on the two waters in the West Country (UK), Blagdon and Chew, as well as on other major stillwater fisheries.

Hook 12-14.
Thread Orange.
Tail None.
Body Orange seal's fur or floss.
Rib Oval gold tinsel.
Hackle Light furnace cock.
Wing None.

Stick Fly

This nymphal pattern is extremely popular with many of Britain's stillwater fly-fishermen. It is thought to imitate a caddis in its case. Some dressers tie the fly up with the attractive fluo floss band at the rear, not in the thorax position.

Hook Long shank 8-10.
Thread Black.
Tail None.
Body Cock pheasant tail fibres.
Rib Gold wire.
Hackle Short natural red hen.
Wing None.
Thorax Fluorescent silk – yellow, green, or orange.

Green Beast

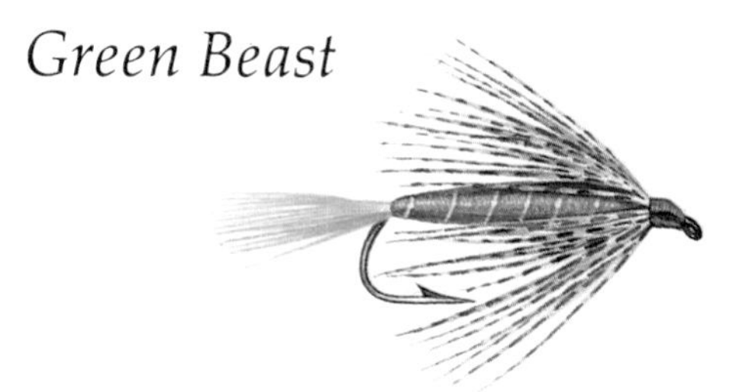

A large nymph devised with aquatic beetle larvae in mind, this fly was devised by Alan Pearson, one of England's best known small-water anglers. His record of catching big fish cannot be bettered.

Hook Long shank 8.
Thread Green.
Tail Green cock hackle fibres tied short.
Body Grass green floss silk tied carrot shape.
Rib Fine silver wire.
Hackle Brown partridge tied sparse.
Wing None.

Clipped Coachman

This fly is an excellent fish seeker. When, on first viewing a stream, there is no obvious surface activity, the Clipped Coachman will usually find a trout or two. The pattern consists of the traditional wet Coachman with a very much abbreviated wing. This fly has been one of my most consistently successful flies on rough streams, taking both grayling and brown trout. It has also worked well on some of the smaller stillwaters.

Hook 12-14.
Thread Black.
Tail None.
Body Peacock herl (can be weighted with copper wire).
Rib None.
Hackle Sparse brown hen.
Wing None as such, but a short tuft of white feather fibre protudes from the head.

Persuader

Though termed a lure by some anglers, this fly of John Goddard's works well as a sedge/caddis imitation in smaller sizes. Its success is probably associated with its white brightness; the fish can see it, and that makes all the difference.

Hook 8-12 (or long shank 8-10).
Thread Orange or brown.
Tail None.
Body White ostrich herl.
Rib Round silver tinsel.
Hackle None.
Wing case Dark brown turkey herl.
Thorax Orange seal's fur.

Wonder Bug

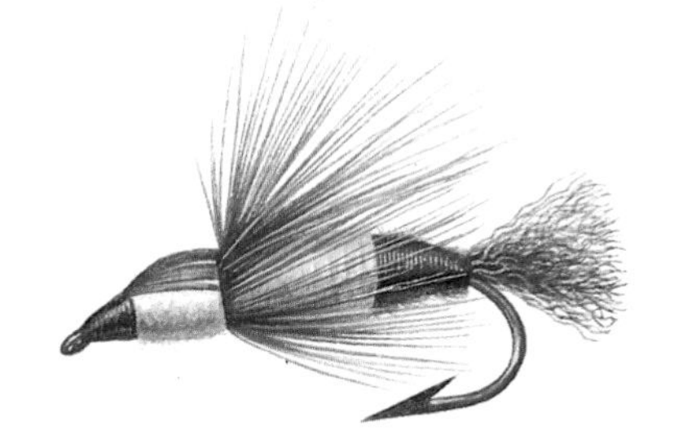

This pattern was created by Alan Pearson for big fish on the clear put-and-take fisheries such as Avington. There are a number of flies called Wonder Bug, but Pearson's is one of the best. Like most of his flies, it is heavily weighted.

Hook Long shank 8-10.
Thread Black or brown.
Tail Brown cock hackle fibres or teased out floss.
Body One-third brown floss; two-thirds pale yellow wool (can be weighted).
Rib None.
Hackle Natural red hen tied in before the thorax.
Wing case Cock pheasant tail fibres.
Thorax Light yellow wool.

PVC Nymph

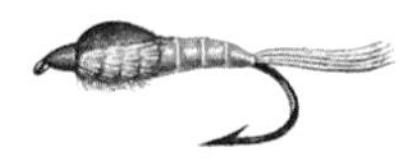

Another pattern from John Goddard, this small nymph with a PVC overbody represents a number of small olive nymphs, the pond olive in particular. The plastic overbody gives a natural effect under water, resembling a real, translucent nymph.

Hook 12-16.
Thread Brown.
Tail Olive condor herl tips.
Body Underbody of copper wire; olive condor herl over this, followed by a wrap of clear PVC.
Rib None.
Hackle None.
Wing case Dark cock pheasant tail fibres.
Thorax Olive condor herl.

Cockwill's Lead Nymph

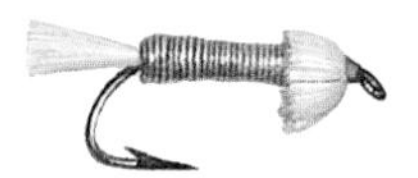

Peter Cockwill has caught so many large fish in his career that it does not bear thinking about. This particular simple nymph is used only on clear waters where you can see your fish. The fly is cast in the fish's path, allowed to sink, then raised to induce the take.

Hook 8-10.
Thread Black or olive.
Tail Olive floss.
Body Close turns of lead wire.
Rib None.
Hackle/Legs Olive floss.
Wing case Olive floss.
Thorax Olive floss.

Long-tail Damsel

When disturbed the natural damsel nymph swims with a very swift, undulating body movement. This fly of Pete Cockwill's is tied to emulate the natural creature's movement. It is usually fished on a slow sink line, and with an exceptionally fast retrieve.

Hook Long shank 10-8 or standard size 8-6 (weighted).
Thread Black or olive.
Tail Olive marabou.
Body Black and olive chenille.
Rib None.
Hackle Palmered olive cock.
Wing None.

Cockwill's Red Brown

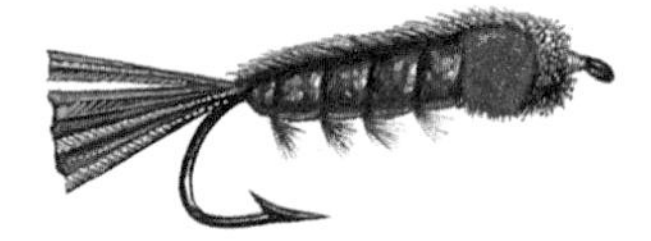

This pattern could be classed as a lure. It is not a nymph in the accepted sense, but it is fished as a nymph-type fly right down on the bottom for those trout that grub down in the silt and weeds. It also makes a very good imitation of a small stickleback. The pattern was based on the early flies devised by the pioneer of modern stillwater angling, Tom Ivens.

Hook Long shank 10-8 (weighted).
Thread Black.
Tail Peacock herl.
Body Bronze Goldfingering.
Rib Peacock herl.
Hackle None.
Back Peacock herl.
Thorax Fluorescent magenta chenille.
Head Peacock herl.

Tellico Nymph

One of America's standard popular nymph patterns. I believe it originally came from the southern states. The original fly was a lot drabber in colour. It works well on rivers and stillwaters.

Hook 10-14.
Thread Black.
Tail Guinea fowl hackle fibres.
Body Yellow floss (can be weighted).
Rib Peacock herl.
Hackle Brown hen.
Back Cock pheasant tail fibres.

Montana Nymph

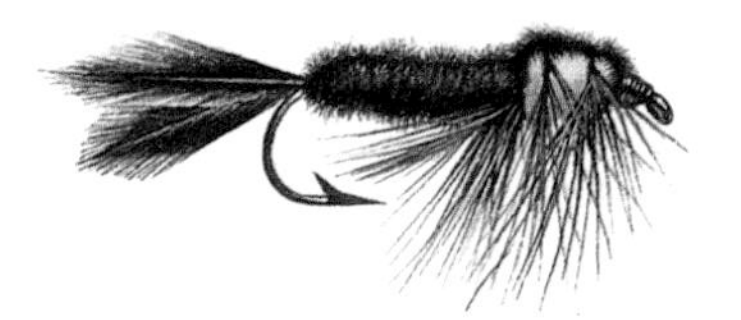

What a good fly this pattern is. Big, black and bold, it was originally tied to represent the creeper (larva of a stonefly) of the black willow stonefly. As its name indicates, it hails from the state of Montana, USA, where it is used on the Yellowstone and Missouri rivers. In the UK it has proved a good fly for reservoir use.

Hook Large shank 6-10.
Thread Black.
Tail Two black hackle tips.
Body Black chenille.
Rib None.
Hackle Black cock hackle wound through the thorax.
Wing case Black chenille.
Thorax Yellow chenille.

Leadwing Coachman Nymph

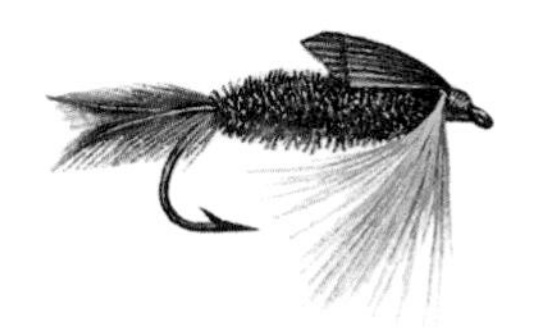

The original Coachman wet fly had a nice, bright white wing. An American variation of this fly sported a wing of grey duck and was called the Lead-wing Coachman. The nymph version of this fly dispenses with the wing as such and uses the grey colour as a wing case or pad.

Hook 10-14.
Thread Black or brown.
Tail Brown hackle fibres.
Body Peacock herl.
Rib None.
Hackle Brown hen fibres.
Wing case Grey mallard wing fibres.

American Gold-ribbed Hare's Ear Nymph

The American version of the British GRHE with the addition of a black feather wingcase. This fly catches fish on both sides of the Atlantic.

Hook 10-14.
Thread Black or brown.
Tail Hare body fur.
Body Hare fur.
Rib Gold wire.
Hackle Hare fur fibres picked out.
Wing case Black feather fibre.
Thorax Hare fur.

Swannundaze Stonefly Creeper

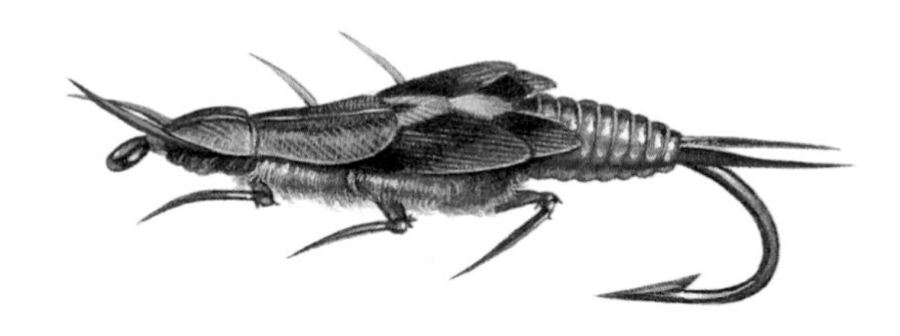

This fly was devised by Frank Johnson of Lyndhurst, New Jersey, using the versatile Swannundaze nymph body material which he markets throughout the world. The fly was tested on such rivers as the Beaverkill and Delaware and it was not found wanting.

Hook Long shank 12-6.
Thread Black or brown.
Tail Brown goose biots.
Underbody Built up with two strips of heavy nylon then covered with amber fur.
Overbody Translucent brown Swannundaze nymph material.
Rib Dark brown thread.
Thorax Amber fur.
Wing case Cock pheasant church window feathers.
Legs Brown partridge.
Antennae Goose biots.

Light Cahill Nymph

A standard American nymph pattern tied to represent the larval form of many of the indigenous mayflies in the rivers of the Eastern States.

Hook 10-18.
Thread Cream.
Tail Wood duck fibres.
Body Creamy tan fur.
Rib None.
Hackle Few fibres of wood duck.
Wing case Wood duck fibres.
Thorax Creamy tan fur.

Blue-winged Olive Nymph

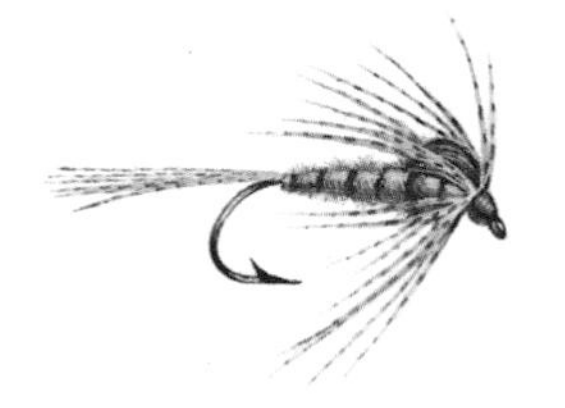

This nymph, as its name suggests, imitates the larval form of the blue-winged olive in the USA, such species as the *Ephemerella cornuta* and others, as well as the British species *E. ignita*.

Hook 14-18.
Thread Olive.
Tail Wood duck fibres.
Body Medium olive fur.
Rib Brown silk.
Hackle Brown partridge fibres.
Wing case Grey goosè wing fibres.
Thorax Medium olive fur.

Dark Hendrikson Nymph

Another standard American nymph pattern devised to imitate a number of different species.

Hook 10-14.
Thread Olive.
Tail Wood duck fibres.
Body Grey brown fur.
Rib Brown silk.
Hackle Brown partridge fibres.
Wing case Turkey tail fibres.
Thorax Grey brown fur.

Olive Tarcher Nymph

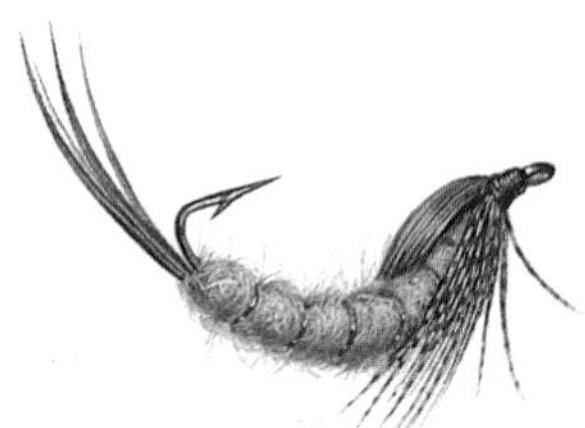

A modern American nymph tied by Ken Iwamasa, an art professor at the University of Colorado. His 'Tarcher' series of nymphs are tied on the strangely-shaped Mustad 37160 hooks known as Old English Bait hooks. The flies are dressed upside-down using the bend of the hook to give a realistic uptilt to the fly.

Hook Mustad 37160 Old English Bait.
Thread Olive.
Tail Cock pheasant tail fibres.
Body Olive antron dubbing.
Rib Fine oval gold tinsel.
Hackle Brown speckled hen or brown partridge fibres.
Wing case Speckled hen fibres.
Thorax Olive antron dubbing.

Zug Bug

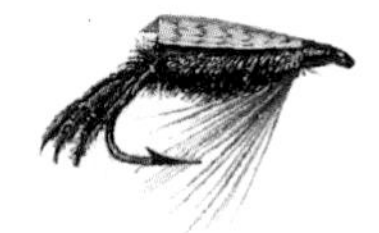

This fly appears in most lists from American fly suppliers. One of the original names for this fly was Kemp's Bug. The pattern is classed as a fancy nymph pattern.

Hook 10-14.
Thread Black.
Tail Peacock sword fibres.
Body Bronze peacock herl.
Rib Flat silver tinsel.
Hackle Brown hen, sparse.
Wing case Wood duck tied at the head (sometimes barred teal is used).

Prince

This fly, like the Zug Bug and Tellico nymphs, is another popular American broad-spectrum nymph. The white strips resemble the clipped-wing Coachman nymph described earlier.

Hook 8-12.
Thread Black.
Tail Brown goose biot (sometimes black goose biot as alternative).
Body Peacock herl.
Rib Flat gold tinsel.
Hackle Collar hackle sparse brown hen (or black).
Wing None.
Horns Two slips of white swan or goose.

Ted's Stonefly

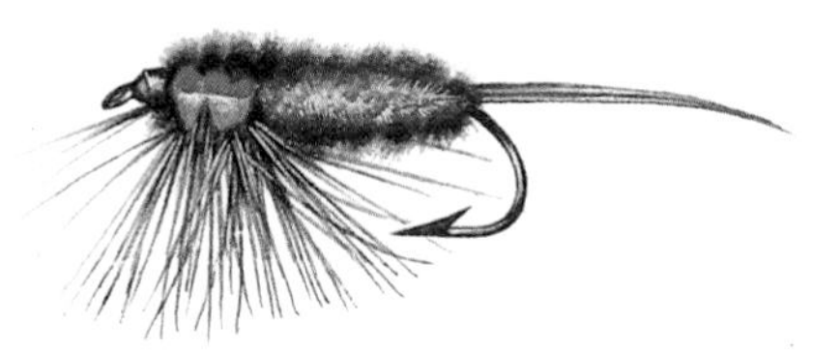

This pattern is a colour variation of the Montana Stonefly nymph given earlier, but in less contrasting colours. The fly was created by Ted Trueblood.

Hook Long shank 6-12.
Thread Black.
Tail Brown goose biots.
Body Brown chenille.
Rib None.
Hackle Black palmered through the thorax.
Wing case Brown chenille.
Thorax Orange chenille.

Hare Caddis

This pattern is used on a fly tied by Darrel Martin of Tacoma, Washington State, USA. It represents a caddis larva in its case. The dubbed fur body gives plenty of movement to the fly.

Hook Shrimp hook, or caddis, or Mustad 37160 8-12.
Thread Black.
Tail None.
Body Dubbed hare fur.
Rib None.
Hackle Sparse brown partridge.
Wing None.
Thorax Yellow floss.
Head Black poly fur with burnt nylon for eyes.

Whitlock Stonefly Creeper

Dave Whitlock is without doubt America's leading fly-dressing innovator and angling artist. His patterns are admired all over the fly-fishing world. He is quick to adapt both traditional and up-to-the-minute materials. His flies have influenced fly-tiers throughout the USA and Europe.

Hook Long shank 2-8.
Thread Orange.
Tail and antennae Boar hair.
Body Mixed amber and orange synthetic dubbing.
Rib Copper or gold wire.
Legs Church window pheasant feather dyed light gold.
Back, wing cases and back of head Brown raffene (Swiss straw).

Teeny Nymph

A simple nymph from the West Coast, USA, that has developed almost into a cult fly. The fly was first tied and named after Jim Teeny of Portland. It is used for all species of trout, and also for steelhead.

Hook Long shank 8-12 (also on short shank).
Thread Black or brown.
Tail None.
Body Cock pheasant tail fibres.
Rib None.
Legs Two tufts of cock pheasant fibre points.
Wing None.

Carey Special

This fly was first created in 1925 by Dr Lloyd A. Day and Col Carey. Its original name was Monkey-faced Louise. The body can be made with a number of different-coloured chenilles, as well as with peacock herl. It was intended to represent a hatching caddis.

Hook 8-12.
Thread Black or brown.
Tail None.
Body Peacock herl.
Rib Black silk.
Hackle Three brown pheasant rump feathers.
Wing None.

Fledermaus

This large nymph is very effective when used as a dragonfly imitation. The fly was first tied by Jack Schneider of San Francisco, and was supposed to emulate – of all things – a bat. This pattern has proved most effective when weighted with an underbody of lead wire.

Hook 8-14.
Thread Brown.
Tail None.
Body Australian possum.
Rib None.
Hackle Collar or pine squirrel.
Wing Barred teal dyed brown (some dressings call for grey squirrel).

Big Hole Demon

Tied by one of the top fly-tyers of the United States, Dan Bailey of Livingstone, Montana. The pattern was tied for the Big Hole River in the State of Montana.

Hook Long shank 9-10.
Thread Black.
Tail Badger hackle fibres.
Body Rear half silver tinsel; front half black chenille.
Rib None.
Hackle Palmered through the black chenille, badger.
Wing None.

Casual Dress

What a nice, shaggy nymph this is, with plenty of action in the water. The Casual Dress was created by one of America's leading fly-dressers, Polly Rosborough. It works well as a dragonfly larva imitation.

Hook Long shank 6-12.
Thread Black.
Tail Muskrat guard hairs and underfur.
Body Muskrat dubbing (can be weighted).
Rib None.
Hackle Collar of muskrat guard hairs.
Wing None.
Head Black ostrich herl.

Martin's Monster

A large American nymph, tied to imitate the nymph of a dragonfly. The body is woven from two shades of chenille. Darrel Martin, its creator, uses strips of soft leather as an under-body to create the real insect shape.

Hook Long shank 8-6 (this can have a bend put in the shank if desired).
Thread Brown.
Tail None.
Body Woven chenille, contrasting olives (other colour combinations if desired).
Rib None.
Legs Two bunches of knotted cock pheasant tail fibres.
Head Dubbed turkey feather flue.
Eyes Bead chain sprayed black.

Green Mountain Damsel

There are many damsel and dragonfly nymph imitations. This one, originated by Randall Kaufmann, is as good as any. This fly is popular throughout America and Canada, and should prove successful anywhere where the needle-like damsel is to be found flitting above the water.

Hook Long shank 8-4.
Thread Olive.
Tail Strands of olive goose biot.
Body Olive green seal's fur.
Back Olive marabou.
Rib Gold wire.
Hackle None, but the fur is picked out to simulate legs.
Wing A tuft of olive marabou, approximately one-third of the body.

Half Back

This is a simple nymph of the broad spectrum variety. It imitates nothing in particular, yet has an appeal to the trout as it looks like food.

Hook 12-8.
Thread Olive or black.
Tail Brown hackle fibres.
Body Peacock herl (thorax also peacock herl).
Rib Gold wire.
Hackle Brown hackle fibres.
Wing Small bunch of brown partridge hackle fibres.

Bitch Creek Nymph

This pattern is a noted big-fish pattern for both trout and Arctic grayling. It is often pre-weighted with lead wire in order to fish it down on the bottom.

Hook Long shank 8-6.
Thread Black.
Tail Two strands of white rubber.
Body Black chenille with a strand of orange chenille under secured by the tying thread. Alternatively, the body can be woven.
Rib None.
Thorax Black chenille.
Hackle Three or four turns of brown hackle palmered through the thorax.
Wing None.
Feelers Two strands of white rubber.

Hot Cat

A South African fly developed by Jack Blackman. He gave me a number of his flies when he came over to this country a few years ago, and I can vouch for the effectiveness of the Hot Cat, which can best be described as a South African version of the Gold-ribbed Hare's Ear.

Hook 6-10.
Thread Black or brown.
Tail Red squirrel tail.
Body Rabbit fur well picked out.
Rib Oval gold tinsel.
Hackle Two to three turns of hot orange cock.
Wing None.

Walker's Nymph

This South African pattern was created by Lionel Walker of Walker's Killer fame. This nymph can come in a variety of colours: green, red and yellow. The first time I tried one of these nymphs, on a local lake, I took a 4 lb (1.8 kg) fighting rainbow on the very first cast.

Hook 8-12.
Thread Red.
Tail Black cock hackle fibres.
Body Red chenille.
Rib Flat gold tinsel.
Hackle Black cock hackle.
Wing case Two brown partridge hackles flanking the body to envelop half of it.

Small's Green Nymph

This South African nymph represents a large number of aquatic creatures, such as small damsel fly larvae. It will take fish anywhere.

Hook Long shank straight eye 10.
Thread Olive.
Tail Partridge tail, two or three strands.
Body Tapered olive green wool (can be weighted).
Rib Oval gold tinsel.
Hackle Sparse natural red hackle.
Wing None.

Elidon Shrimp

Every country has its own shrimp patterns and Australia is no exception. I do not know how effective this fly is, but at least, with its claret colour, this one is a little different.

Hook 8-10.
Thread Black.
Tail None.
Body Rear half gold tinsel; front half dark claret wool.
Rib None.
Hackle Palmered natural red over gold body; black at the body joint, leaving the claret wool unhackled.
Wing None.

Hair Fibre Nymph

An Australian nymph with a very traditional appearance. It would serve well as a nymph for the March brown.

Hook 10-14.
Thread Brown.
Tail Partridge tail fibres.
Body Lightly dubbed blue rabbit underfur.
Rib Oval silver tinsel.
Thorax Hare fur picked out at the sides to resemble legs.
Wing case Partridge tail.

Hare and Copper Nymph

I have used this fly on both rivers and still waters in this country. It can best be described as a New Zealand version of the old favourite, the Gold-ribbed Hare's Ear. It is best fished on a floating line with a long leader.

Hook 10-14.
Thread Brown or orange.
Tail Hare fur guard hairs.
Body Hare fur (weighted).
Rib Copper wire.
Hackle None.
Wing Hare fur picked out.

Horn Caddis

This is a New Zealand pattern representing the caddis larva complete with a case. This fly can be used anywhere in the world. Fish the fly close to the bottom for the best results.

Hook 10-14.
Thread Black.
Tail None.
Body Grey wool tied around the bend (lacquered after the rib).
Rib Silver wire.
Hackle Sparse grizzle to resemble the legs. Narrow band of white wool to resemble the body emerging from the case.

Governor Nymph

Sometimes called the Red-tipped Governor, this New Zealand fly is a nymph version of the traditional Governor wet or dry fly pattern. It comes under the category of a fancy attractor nymph.

Hook 10-12.
Thread Black.
Tail Red hackle fibres.
Body Red silk.
Rib None.
Hackle Natural red hen.
Wing case Hen pheasant.
Thorax Peacock herl.

Orange Nymph

This nymph is tied to represent the New Zealand species of mayfly *Zephlebia cruentata*, called the pepper-winged olive.

Hook 12-14.
Thread Brown or orange.
Tail Short brown partridge fibres.
Body Orange wool.
Wing case Brown partridge.
Rib Fine silver wire.

Precieuse

A French nymph from the Guy Plas stable representing a wide range of Ephemerid nymphs. This pattern comes in a variety of subtle body and hackle colour combinations, and is suitable for all types of water.

Hook 12-18.
Thread Grey or black.
Tail Ash grey hackle fibres.
Body Light grey poly dubbing.
Rib Silver tinsel.
Hackle Short fibred ash grey.
Thorax Dark grey.

Phryga-Nymph

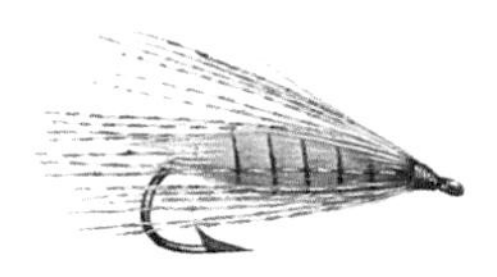

A Guy Plas nymph imitating a hatching sedge/ caddis fly. Like the Precieuse, it can have a number of different-coloured bodies.

Hook 8-10 (short shank or long shank).
Thread Yellow.
Tail None.
Body Yellow and light brown mixed dubbing, tapered towards the head.
Rib Brown silk.
Hackle Ash grey short, at the top of the head.
Side emergent wings Mottled brown and grey spade hackle fibres.

La Mue

This is an interesting French fly tied to represent a stillborn nymph with only one wing released from its nymphal shuck. This pattern is useful for both still and running water.

Hook Long shank 16-18.
Thread Brown.
Tail None.
Body Light olive silk.
Rib Brown silk.
Hackle Short, one side only, brown.
Wing One only, emerging from thorax, mottled brown grey.
Wing base Cock pheasant tail fibres.

Gammarus

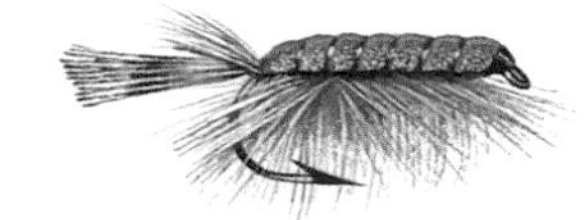

A number of freshwater shrimp patterns have already been illustrated. This Austrian pattern is an example of Roman Moser's fly dressing skills. For most of his flies, Roman Moser utilizes modern, man-made products, skilfully blending them with more traditional materials.

Hook 8-10.
Thread Brown.
Tail Partridge hackle fibres.
Body Body gills (a soft hairy product supplied by some fly-dressing houses).
Rib Copper wire.
Hackle The body gill material provides the hackle.
Back Reflective plastic strip called Spectraflash.

Hydropsyche

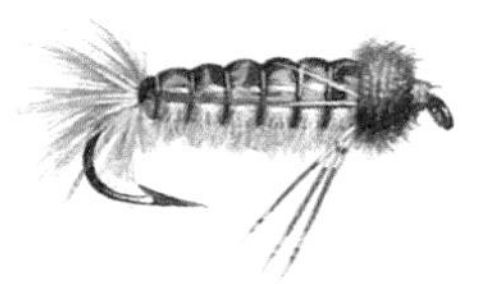

The Hydropsyche species of caddis do not make portable shelters for themselves, unlike many other caddis larvae. Along with Rhyacophila species, they make webs; net-like ones in the case of Hydropsyche, purse-like in the case of Rhyacophila. Both are voracious predators. Both species could be imitated by this pattern of Roman Moser's.

Hook Long shank 10.
Thread Black or brown.
Tail Soft, fluffy feathers from the base of a partridge hackle.
Body Body gill material.
Rib Copper wing.
Hackle Sparse partridge hackle. Latex marked with a dark brown marker pen.
Head Squirrel fur dubbing.

Floating Sedge Pupa

The combination of deer hair and foam makes this Austrian pupal pattern an excellent floater. Grey undyed deer hair is used in one pattern; brown-dyed deer hair can be used as an alternative dressing.

Hook 10.
Thread Brown.
Tail Deer hair.
Body Deer hair.
Rib None.
Hackle None.
Wing None.
Head Thin polyurethane foam.

Ucero

One of the leading angling authorities on the caddis fly in the USA is Gary Lafontaine. His book, *Caddis Flies*, is considered by many to be the definitive work on the subject as far as fishing is concerned. This hatching sedge pattern from Luis Antunez of Madrid is based on Lafontaine's patterns.

Hook 10-12.
Thread Red.
Tail Short tuft of polypropylene yarn.
Body Light brown or cinnamon polypropylene dubbing.
Rib Brown thread.
Hackle Poly yarn flared out, sloping backwards.
Wing None.

Arthofer Nymph

A popular Austrian nymph used in many parts of Europe, created by Louis Arthofer. It can be used on both rivers and stillwaters.

Hook 8-12.
Thread Black or brown.
Tail Three fine strands of ostrich herl dyed brown.
Body Brown ostrich herl.
Thorax Copper wire.
Hackle Guinea fowl or partridge hackle clipped short.
Wing case Hen pheasant wing feather.

Grannom Pupa

This Spanish pupa pattern imitates the pupa of the early hatching sedge, the grannom, once called the greentail (*Brachycentrus subnubilis*). The grannom makes its appearance early in the fishing season. This imitation can also be used to imitate a number of other species, and can be fished on still waters as well as rivers.

Hook 12.
Thread Black or brown.
Tail None.
Body Light brown poly dubbing.
Rib Dark brown thread.
Hackle Short tuft of brown partridge.
Wing Dark grey duck slips.
Head Reddish brown poly dubbing.

Floating Pupa

This tiny fly from Luis Antunez can imitate a wide variety of nondescript creatures found in the surface film, including the hatching micro sedges, and members of the family Diptera.

Hook 16-18.
Thread Green.
Tail None.
Body Bright green poly dubbing or seal's fur.
Rib Brown thread.
Hackle None.
Wing None.

The Dormouse Nymph

In the fruit-growing areas of Yugoslavia, one of the biggest pests and threats to the crops is the dormouse. About 200,000 have to be trapped each year, and some people eat these little mice (shades of ancient Roman cuisine). The skins have no commercial value and are disposed of – except the few that find themselves in the hands of fly-dressers. This nymph, devised by Marjan Fratnik, uses the fur from the dormouse tail. This fur almost breathes in the water when wet.

Hook 10-14 (can be weighted with lead wire).
Thread Black.
Tail Dormouse tail fibres.
Body Fur from a dormouse tail.
Rib None.
Thorax Dormouse tail fur.
Wing case Any dark feather.

Cocchetto Nymph

In the area around Milan people used to collect the silken cocoons of a species of indigenous moth. This primitive raw silk was used to form the body dubbing of nymphs; the underlying body colour would show through the silk when the nymph was wet. According to Luciano Maragni of Milan, this type of fly was banned on some Slovenian rivers as it proved too killing for grayling.

Hook 14-16.
Thread Purple.
Tail None.
Body Moth cocoon dubbing over underbody of tying thread.
Rib Clipped palmered dark natural red hackle.
Hackle None.
Wing None.

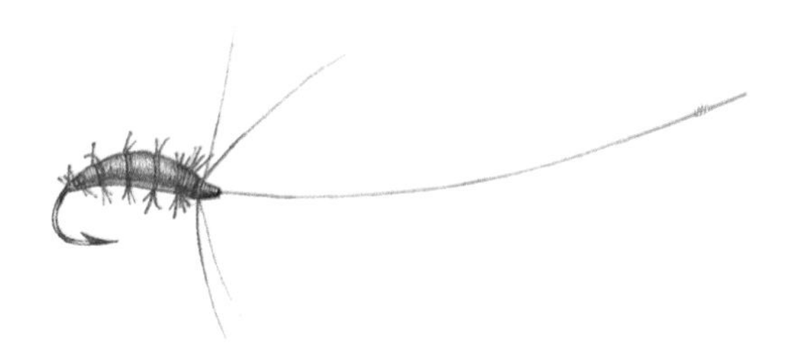

Hook 14-16.
Thread Black.
Tail None.
Body Moth cocoon dubbing.
Rib Clipped dark natural red hackle.
Hackle A few wisps of unclipped hackle.
Wing None.

LURES (streamers and bucktails)

Despite recent developments and increased use of these larger flies, lures are not a new innovation. It is very likely that this form of trailing fly has a lineage as long, if not longer, than the conventional patterns to imitate insect life.

The Eskimos of Alaska have been using lures made of polar bear hair for many years. How long? There is no way of telling, but it could have been for centuries.

Wherever primitive cultures have sought to catch fish with a hook, it is possible that some form of highly active fur or feathered lure could have been used. Who was the first fisherman to put feathers on a hook and catch the ubiquitous mackerel? This simple and effective lure was a steamer fly, plain and unadorned.

The modern streamer/bucktail lure can be traced back to the shores of the USA, and to the pioneer and doyen of the American dry fly, Theodore Gordon himself. He was alleged to have fished for pike with a large, streaming, feathered fly in the late 1880s. He subsequently published the dressing around 1903 and called his unruly, far-from-dry fly, the Bumblepuppy.

Many years before Gordon's canine lure, Irish fishermen sought the large pike of the limestone loughs by means of ultra-large flies tied with whole peacock eye feathers for the wing. However, such flies were more in the salmon fly mould, rather than like the streamer fly of Theodore Gordon.

Although large streamer-type flies (other than the normal salmon flies), such as the multi-hook Terrors and other types of lure, have been used by British anglers, it is to America that we must look for the origin of most modern patterns. A large proportion of British lures are developments or adaptations of American flies, often with no credit given to the origin of the patterns. Some so-called new killing flies are created by tyers whose purpose is to sell the modern materials that they have exclusive rights to. Such flies usually never last; as new products come along they fade into obscurity. Ninety-nine per cent of British lures are used on stillwaters, whilst the opposite holds true in the USA. Most of their large flies have been devised for use on rivers.

Lures are large flies tied to represent indigenous baitfish or small fry of the area they are fished in. Even infant trout and salmon have not escaped the attention of the fly-dresser, for such immature fish are taken by their larger brothers and sisters.

The members of one group of lures are so vibrant and vivid in colour that they could not possibly represent any small fish that swam in any river in North America or Europe. Their nearest natural equivalent may well be found around some tropical coral reef, not in a trout stream or lake. But their purpose is to attract, to make the quarry sit up and take notice. Trout will attack such gaudy flies from anger or curiosity. And if the lure is sufficiently fish-shaped, then that too will prompt the feeding response, even though the colour may be somewhat unatural.

Let me qualify what is meant by streamer and bucktail. Streamers are lures tied with feathered wings, usually cock neck or saddle hackles. Sometimes other types of feather are used, or even a feather and hair combination. Bucktails, on the other hand, use hair as the winging medium. As the name implies, the original flies used hair from various deer tails for the wing. Normally, flies using goat, calf, squirrel and even yak are called bucktails. Modern man-made fibres are sometimes used in some patterns.

All species of game fish are taken on streamers and bucktails: brown trout, rainbows, dolly varden, cut-throat, migratory seatrout, steelhead, Atlantic salmon and various species of Pacific salmon, pike and other predatory coarse fish, and a whole range of saltwater sport fish.

The large fly is not the exclusive province of America and Great Britain, far from it. The unique lures of New Zealand bear witness to this. A number of New Zealand flies are given here, but there are many hundreds more. The lures of New Zealand have influenced fly design in both Australia and South Africa.

The two definitive works on the subject of streamers and bucktails are American: *Streamer Fly Tying and Fishing* by Joseph D. Bates (Stackpole), and *Fishing Flies and Fly Tying* by William Blades (Stackpole). Both books were published in the early 1950s. However, most fly-tying books include various lure patterns, for this type of fly has become an important addition to the modern fly fisherman's armoury.

Muddler Minnow

This is a particular favourite of mine. It can be used in many guises: as a small fish; a large nymph; on the surface it can be used as a caddis fly or stonefly; and it can be used to imitate grass-hoppers and crickets. The fly was created by Don Gapen of Anoka, Minnesota, and was used on the River Nipogen as an imitation of a small baitfish of the Gobio species, called by the local Indians the Cockatush and by others muddler. Since its inception, many varieties and adaptations have been created, some of which are here.

Hook All sizes long shank 4-10.
Thread Black or brown.
Tail Oak turkey slip.
Body Gold tinsel.
Rib Oval gold tinsel (optional).
Hackle Collar of unclipped deer hair.
Wing Grey squirrel with oak turkey slips either side.
Head Clipped deer hair.

Orange Muddler

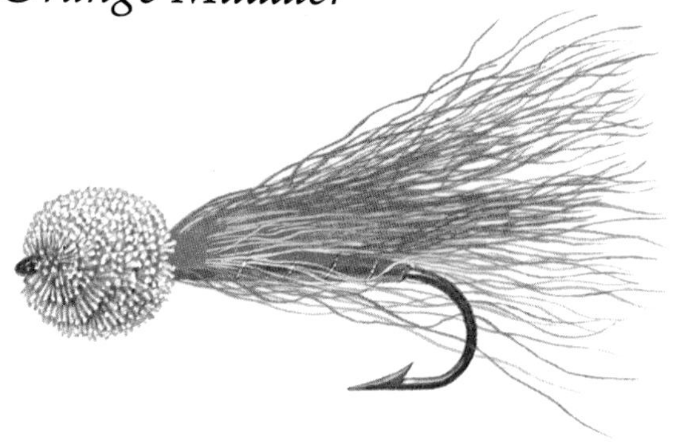

This is a variation on the standard Muddler. I published the dressing in *Lures* in the early 1970s. It is useful when trout ignore the more sombre standard pattern.

Hook Long shank 6-10.
Thread Orange.
Tail Orange feather fibre (optional).
Body Gold tinsel.
Rib Oval gold.
Hackle Unclipped deer hair.
Wing Orange squirrel or bucktail.
Head Clipped deer hair.

Black Muddler

As with the Orange Muddler, the black version followed very quickly on the tail of the original fly, and is used when a dark fly is required. There is also an all-white Muddler.

Hook Long shank 6-10.
Thread Black.
Tail None.
Body Black silk.
Rib Silver oval.
Hackle Unclipped deer hair collar.
Wing Black bucktail or other hair.

Whiskey Fly

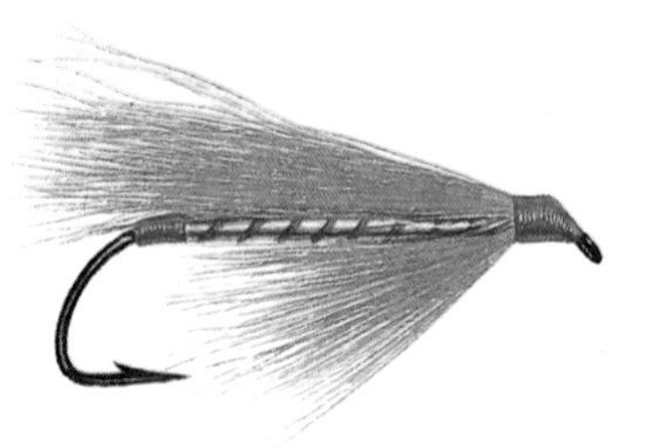

A British bucktail lure devised by Albert Whillock and used originally on Hanningfield reservoir in Essex. The original fly utilized a proprietary brand of self-adhesive silver tape on the body. It is more usual now to use the dressing given here.

Hook Long shank 6-10.
Thread Orange.
Tail None.
Body Flat gold tinsel.
Rib Fluorescent red floss.
Hackle Hot orange cock hackle.
Wing Orange bucktail or calf tail.

Sweeney Todd

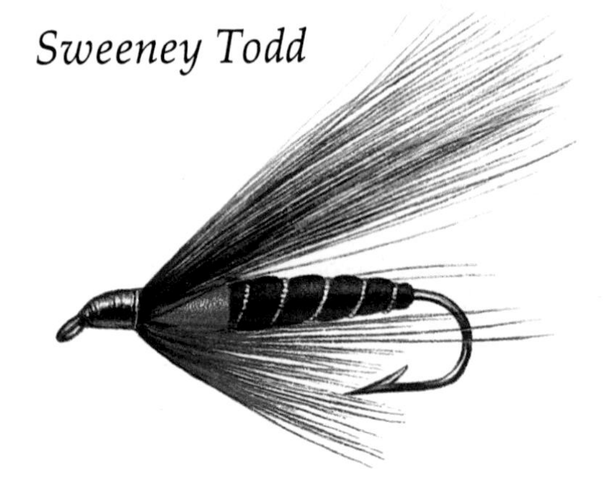

This demon barber of a fly is perhaps one of the most famous of the late Richard Walker's flies. A very killing 'black lure' fly; the addition of the fluorescent throat enhances the pattern.

Hook All sizes long shank 6-10; also tandem.
Thread Black.
Tail None.
Body Black floss, with fluorescent magenta wool at throat.
Rib Fine oval silver tinsel.
Hackle crimson.
Wing Black squirrel.

Lead Head

This fly is the fore-runner of the much used Dog Nobbler. The weight is at the head of the fly, which gives it a diving action that attracts the trout, but makes it rather difficult to cast. It serves well as a jigging lure.

Hook Long shank 6-10.
Thread Black or brown.
Tail None.
Body Black (but other colours also useful).
Rib Oval silver tinsel.
Hackle None.
Wing Black squirrel.
Head Lead shot painted black, with the addition of an eye.

Dog Nobbler

One of the most recent killing lures on British reservoirs. It was devised by Trevor Housby, the well-known angling all-rounder. This fly has the rather dubious distinction of being one of the few flies to have its name registered. The fly itself is a development of American jigging lures. An almost identical fly was tied by the American, Bill Blades, and called Ice Fishing Fly. It was published in his book *Fishing Flies and Fly Tying* in 1951.

Hook Long shank 6-10.
Thread Black.
Tail Black marabou (other colours: orange, white, pink, yellow, etc.)
Body Black chenille (or other colours to match tail).
Rib Oval silver tinsel (optional).
Hackle Sometimes a collar hackle is added to match body and tail colour.
Wing None.
Head Lead shot painted with eye if desired.

Baby Doll

A very simple fly which imitates small fry. The first recorded use of such flies was on the Midlands reservoir, Ravensthorpe, in the early 1970s. Its originator was Brian Kench. Since its creation there have been many colour variations. An all-lime-green version in fluorescent wool has proved most effective; so has a peach-coloured one for anglers on Grafham Water. Combination dolls, that is, with different coloured backs to the body, are also used to some effect.

Hook Long shank 6-14.
Thread Black.
Tail White wool.
Body White wool.
Rib None.
Hackle None.
Back White wool.

Appetiser

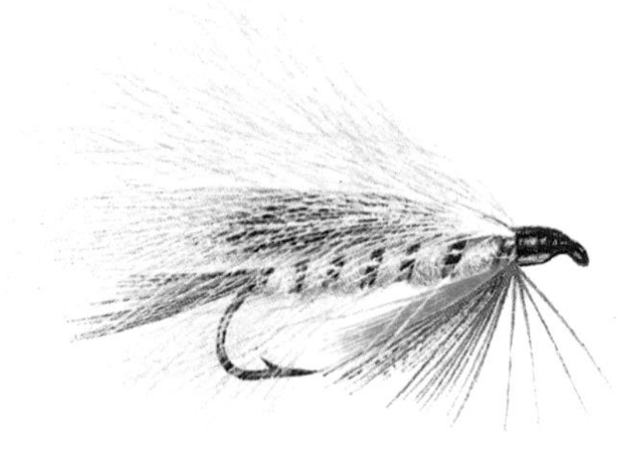

A fly devised by Bob Church, one of Britain's leading stillwater anglers. This fly imitates the fry of such coarse fish as the roach. It can, on its day, be a very killing pattern indeed.

Hook Long shank 6-10.
Thread Black.
Tail Mixed orange, green and silver mallard fibres.
Body White chenille.
Rib Oval silver tinsel.
Hackle As tail.
Wing White marabou with grey squirrel over.

Missionary

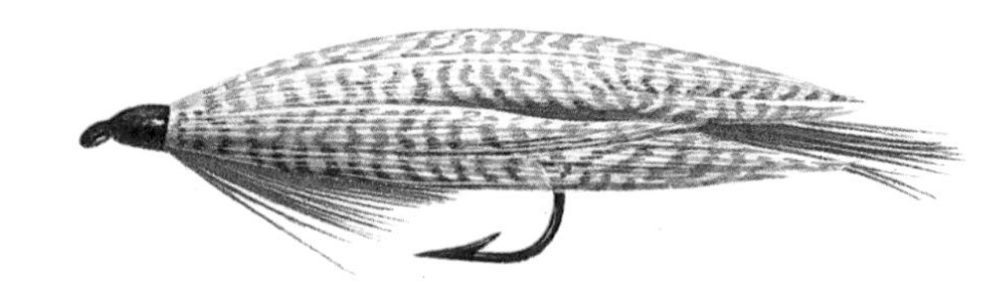

This pattern is a modern reservoir-lure version of a much older wet fly, the dressing of which was published in Courtney Williams's *Dictionary of Trout Flies*. He credits the invention of the original Missionary to Captain J. J. Dunn, and it was used on Blagdon Water. Williams also gives an orange-hackled version. The original flies were developed by Dick Shrive, another Midlands reservoir stalwart, into the lure given here.

Hook Long shank 6-10.
Thread Black.
Tail Red cock hackle fibres.
Body White chenille.
Rib Flat silver tinsel.
Hackle Red cock hackle.
Wing Grey mallard breast feather tied flat on top of the hook.

Black and Orange Marabou

The first time I used this fly, I caught two fish on successive casts. Both fish were about 3 lb (1.5 kg) in weight, which in those days were classed as very good fish, and on that day no other fly seemed to work. The Black and Orange Marabou has continued to catch fish wherever I have used it.

Hook Long shank 8.
Thread Black.
Tail Orange cock hackle fibres.
Body Flat gold tinsel.
Rib Oval gold tinsel.
Hackle Hot orange cock hackle fibres.
Wing Black marabou with Jungle cock cheeks.

Orange Bucktail

This fly can best be described as a fore-runner of the better known Whiskey Fly. I tied this pattern up because there were times when the colour orange was the only thing that the trout on my local waters wanted.

Hook Long shank 6-10.
Thread Orange or black.
Tail None.
Body Large oval tinsel in close turns.
Rib None.
Hackle None.
Wing Orange calf or bucktail.
Head Can be painted with an eye, but this is optional.

Black Shimma

I based this fly on an American West Coast fly called the Black Bugger. I added some strands of a highly shimmering tinsel – which I have called Shimma – to the tail. The fly caught fish in embarrassing numbers at the tail end of the 1984 season, and continued to do so right through 1985 whenever the use of a lure was called for.

Hook Long shank 6-10.
Thread Black.
Tail Large tuft of black marabou with about eight strands of silver Shimma (this is a twisted metallic thread).
Body Black chenille.
Rib None.
Hackle Palmered black cock.
Wing None.

Pink Shimma

A pink version of the previous pattern. Other colours that have proved successful are white, brown and olive.

Hook Long shank 6-10.
Thread Black.
Tail Large tuft of pink marabou along with eight strands of pink Shimma.
Body Pink chenille.
Rib None.
Hackle Palmered pink cock hackle.
Wing None.

Rees's Lure

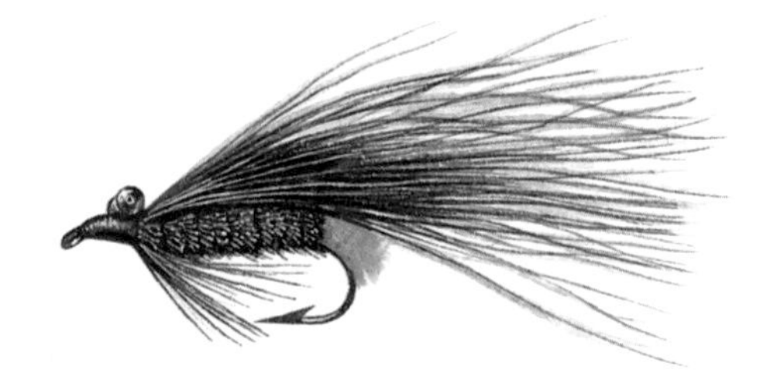

One of the best all-round lures. I have lost count of the number of fish it has caught for me and my friends in many parts of the world. I was first given the fly by Trevor Rees of Cardiff. Other versions are white, orange and yellow.

Hook Long shank 6-10; also on tandem linkage.
Thread Black.
Tail Fluorescent lime green.
Body Bronze peacock herl.
Rib None.
Hackle Black cock hackle fibres.
Wing Black marabou with black squirrel over.
Head Silver bead chain eyes.

Leprechaun

There used to be a lake in the south-east of England that held good-quality brown and rainbow trout. However, it was a water where I never caught any decent fish, while everybody else did well. It reached the stage when I talked myself out of catching anything before I even started to fish, such was my lack of confidence. I mention this because the only fish I ever caught there was on a Leprechaun. This green lure was created by Peter Wood.

Hook Long shank 6-10.
Thread Black or green.
Tail Green cock hackle fibres.
Body Fluorescent lime green chenille.
Rib Flat silver tinsel.
Hackle Green cock hackle fibres.
Wing Four green cock hackles.

Black Chenille

Bob Church's version of the standard Black Lure. This fly is extremely popular on British reservoirs as an early-season lure.

Hook Long shank 6-10.
Thread Black.
Tail Black cock hackle fibres.
Body Black chenille.
Rib Flat silver tinsel.
Hackle Black cock hackle fibres.
Wing Four black cock hackles.

Marabou Muddlers

The original Muddler travelled westwards and was adapted by Dan Bailey of Montana, USA. He enhanced an already-effected pattern with a mobile marabou feather hackle. The white and orange versions are depicted here. Other flies in the series are black, yellow, brown and green.

Hook Long shank 4-8.
Thread Black.
Tail Red cock hackle fibres.
Body Silver tinsel chenille.
Rib None.
Hackle Collar of unclipped deer hair.
Wing White marabou with three or four strands of peacock sword.
Head White clipped deer hair.

Hook Long shank 4-8.
Thread Black.
Tail Red cock hackle fibres.
Body Gold tinsel chenille.
Rib None.
Hackle Unclipped deer hair.
Wing Orange marabou and three or four strands of peacock sword.
Head Clipped deer hair.

Ace of Spades

A very killing fly created by Dave Collyer of Surrey, UK. This fly is tied in the New Zealand Matuka style, with the wing bound down by the ribbing tinsel. This fly differs from the normal Matuka flies in the fact that Collyer has overlaid the black main wing with a veiling of brown mallard. Why this makes a difference I do not know, but without this brown addition, the fly does not appear to be as killing.

Hook 6-12.
Thread Black.
Tail None.
Body Black chenille.
Rib Oval silver tinsel.
Hackle Guinea fowl hackle fibres.
Wing Tied matuka style with four black cock hackles, with bronze mallard over.

Orange Streamer

I devised this attractor lure, with its stripes in the wing, as a streamer pattern to imitate a perch fry. It is quite effective as a general orange lure pattern.

Hook Long shank 6-10.
Thread Black or orange.
Tail None.
Body Orange floss.
Rib Oval gold tinsel.
Hackle None.
Wing Two orange cock hackles flanked by two grizzle cock hackles. Jungle cock cheeks either side optional.

Jack Frost

Another fly created by Bob Church, this white lure is at its best at the tail end of the season when the big trout are chasing fry.

Hook Long shank 6-10.
Thread White or black.
Tail Crimson wool.
Body White wool covered with a strip of wound polythene.
Rib None.
Hackle Crimson cock hackle with a white cock hackle in front.
Wing White marabou.

Undertaker

A black version of the Baby Doll. The adornment of a silver rib adds a little flash and glitter to brighten up what would otherwise be a very mournful fly.

Hook Long shank 8-10.
Thread Black.
Tail Black wool.
Body Black wool.
Rib Fine silver tinsel.
Hackle None.
Wing None.
Back Black wool.

Yellow Chenille

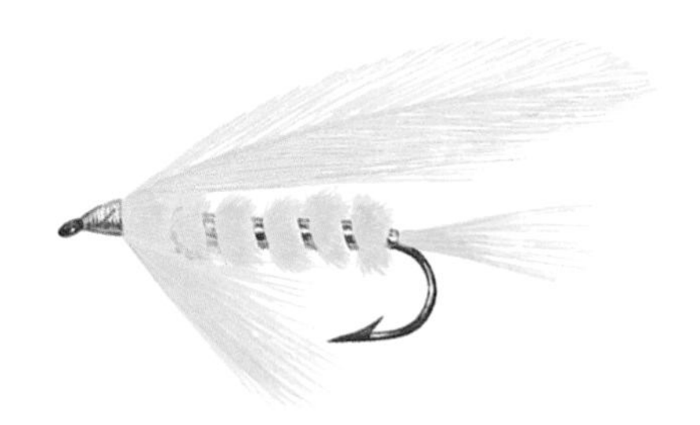

This is a brighter version of the Black Chenille. Like the Goldie, it is a good fly when the water is coloured.

Hook Long shank 6-10.
Thread Black or yellow.
Tail Yellow cock hackle fibres.
Body Yellow chenille.
Rib Flat silver tinsel.
Hackle Yellow cock hackle fibres.
Wing Four yellow cock hackles.

Polystickle

This pattern, devised by Richard Walker, owes a lot to an earlier fly invented by Ken Sinfoil when he was in charge of Weirwood reservoir in Sussex. He called his pattern Sinfoils Fry. In its construction, in order to give the fly a transparent effect like the real fry it was imitating, Ken used a strip of polythene wound for the body. Richard Walker adapted this polythene body and by using raffene (Swiss straw) for the back, he created a series of Stickle flies in various colours to imitate the immature fish of different species. This type of fly has also worked quite well for seatrout. The best polythene to use is gauge 250. The raffene must be dampened before stretching it over the fly's back.

Hook Long shank 8-10 (can be silvered or nickle hook).
Thread Black or brown.
Tail Raffene (Swiss straw).
Body Build up to a fish shape with a polythene strip; an underbody of red floss or wool can be added at the throat.
Hackle Cock hackle to match the raffene.
Back Raffene.
Combinations Black raffene and scarlet hackle; brown raffene and orange hackle; green raffene and brown hackle.

Jersey Herd

Another fry-imitating fly, this time created by Tom Ivens, one of the major influences on modern stillwater fishing. The name Jersey Herd came from the fact that the tinsel on the original fly was cut from a milk bottle top from a bottle of Jersey milk. An equally effective body can be made from the yarn, Goldfingering, in a bronze colour.

Hook Long shank 8-10.
Thread Black.
Tail Peacock herl.
Body Copper flat tinsel over an underbody of silk.
Rib None.
Hackle Collar hackle of hot orange.
Wing None.
Back Bronze peacock herl.
Head Bronze peacock herl.

Viva

A very popular fly on some waters, this is a simple black lure with a fluorescent green tail. The colour fluo green is extremely attractive to the trout. It is by far and away the most effective of the fluorescent range of colours. Alternative winging mediums for this pattern are black squirrel, black goat, or black marabou, and sometimes a combination of hair and marabou. An equally effective fly is the white version, the White Viva. The tail is the same in both patterns; just change white for black for the rest of the fly.

Hook Long shank 6-10.
Thread Black.
Tail Green fluorescent wool.
Body Black chenille.
Rib Flat silver tinsel.
Hackle Black cock hackle fibres.
Wing Four black cock hackles (some prefer a wing of black squirrel).

Goldie Lure

A very good fly for murky water conditions. Yellow is a very good colour for such conditions, and a contrasting yellow/black is even better. This is a favoured pattern for Alan Pearson, holder of the British records for rainbow and brook trout.

Hook Long shank 6-10.
Thread Black.
Tail Yellow hackle fibres.
Body Flat gold tinsel.
Rib Gold wire.
Hackle Yellow cock hackle fibres.
Wing Yellow goat with black goat over.

Alaska Mary Ann

A classic American bucktail fly. This fly was based on a primitive Eskimo ice-fishing lure called the Kobuk Hook, a jig used by the Eskimos of the Kotzebue area of Alaska. Apparently it was made with a sliver of whalebone or walrus ivory, a nail, and some polar bear hair. The fly was adapted by Frank Dufresne in the early 1950s.

Hook Long shank 4-10.
Thread Black.
Tail Red hackle fibres (sometimes scarlet wool).
Body White chenille.
Rib Silver tinsel.
Hackle None.
Wing White calf tail (the original called for polar bear). Jungle cock cheeks optional.

Spruce

One of the most popular West Coast streamer patterns for cut-throat, rainbow, and brown trout. It was combined with the Muddler to form the pattern called the Spuddler.

Hook Long shank 4-10.
Thread Black.
Tail Four or five peacock sword points.
Body One-third red floss silk; the rest thickly tied peacock herl.
Rib None.
Hackle Collar hackle of badger cock.
Wing Two badger hackles usually tied splayed out.

Mickey Finn

Perhaps one of the best-known American streamer or bucktail patterns, this one can be tied either as a hair-wing or as a feathered lure. The original name was Red and Yellow; it was rechristened the Mickey Finn in the late 1930s. It has also been tied up as a salmon fly.

Hook Long shank 6-10.
Thread Black.
Tail None.
Body Embossed silver tinsel (or flat silver tinsel).
Rib If flat tinsel is used for the body then an oval silver rib is required.
Hackle None.
Wing In three parts yellow bucktail divided by a bunch of red bucktail.

Thunder Creek Red Fin

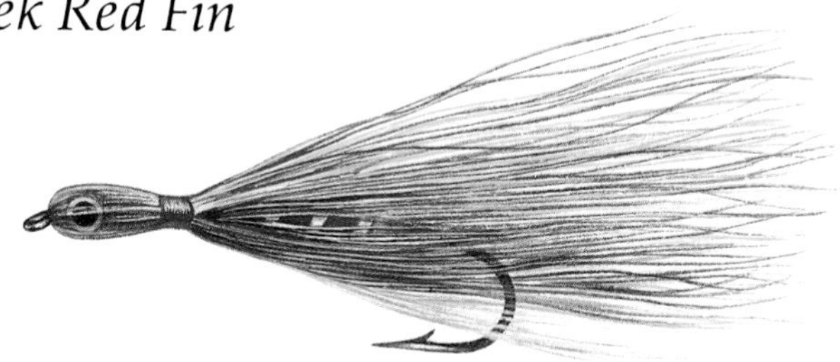

This is one of a series of bucktail lures created by the American fly-dresser, Keith Fulsher, a banker from Eastchester, New York. All of the Thunder Creek series are tied in the same way with the back and wing tied down to form the head of the fly. All the flies imitate a variety of small indigenous baitfish.

Hook Long shank 8-10.
Thread Red.
Tail None.
Body Red floss.
Rib Flat silver tinsel.
Back Brown bucktail.
Belly White bucktail. Bound down to form the head.
Head Yellow eye, black pupil.

Thunder Creek Silver Shiner

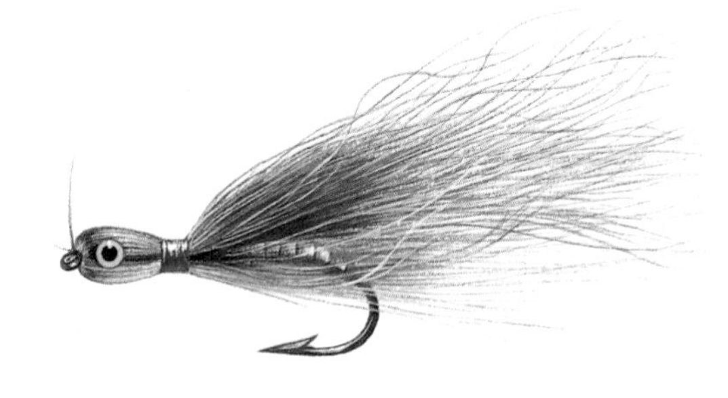

Believed to be the favourite of its originator. Others in the series are, apart from the Red Fin already mentioned, Golden Shiner, Black-nosed Dace, Satin Fin Minnow, Emerald Minnow and the Spot-tailed Minnow.

Hook Long shank 8-10.
Thread Red.
Tail None.
Body Flat silver tinsel.
Rib Oval silver tinsel.
Back Brown bucktail.
Belly White bucktail; bound down to form the head.
Head Yellow eye, black pupil.

Black Ghost

An American streamer fly that has served many fisherman well in this country, myself included. On one occasion this fly won me a charity fly fishing match against the Houses of Parliament team. The water that particular day was so murky that it needed a fly that the fish could see. The Black Ghost was the answer. The contrasting black body and white wing with yellow was highly visible to the fish. The fly originated in Maine.

Hook Long shank 6-10.
Thread Black.
Tail Yellow hackle fibres.
Body Black floss.
Rib Flat silver tinsel.
Hackle Yellow cock hackle fibres.
Wing White cock hackles (Jungle cock cheeks optional).

Badger Streamer

An attractive pattern devised by William F. Blades, a fly-dresser who was way ahead of his time. His book *Fishing, Flies and Fly Tying* was published in 1951. The barred wood duck tail is a feature of many of Bill Blade's flies. Poul Jorgensen, the well-known American tyer, was a student of Blades. Many so-called modern innovations can be found in the pages of Blades's book.

Hook Long shank 6-10.
Thread Black.
Tail Barred wood duck flank.
Body Embossed silver tinsel.
Rib None.
Hackle Beard hackle of white bucktail.
Wing White bucktail, four badger hackles tied over flanked by barred wood duck.
Head Black with white eye and black pupil.

Black-nosed Dace

This pattern was tied by one of America's leading fly-tyers and fishing authors, Art Flick. It works very well on British stillwaters as a general fry imitation.

Hook Long shank 6-10.
Thread Black.
Tail Short tuft of red wool.
Body Embossed silver tinsel (or flat silver tinsel).
Rib None if embossed. If flat silver body, then oval silver tinsel is required.
Hackle None.
Wing In three parts: brown bucktail, over black bear, over white bucktail.

Grey Ghost

Perhaps one of the USA's best-known streamer patterns, it was tied to imitate a smelt baitfish. The originator of this fly was the justly famous Mrs Carrie G. Stevens. The pattern dates from around 1924 and is one of a number of streamer lures credited to this fly-tyer. All of Mrs Stevens's flies were recognizable by a ring of red varnish or tying thread applied to the head of the fly.

Hook Long shank 4-10.
Thread Black.
Tail None.
Body Orange floss.
Rib Flat silver tinsel.
Hackle A few strands of white bucktail, four strands of peacock sword and a golden pheasant crest curving upwards.
Wing Four grey cock hackles over a bunch of white bucktail, flanked by silver pheasant body feather. Jungle cock cheeks optional.

Edson Light Tiger

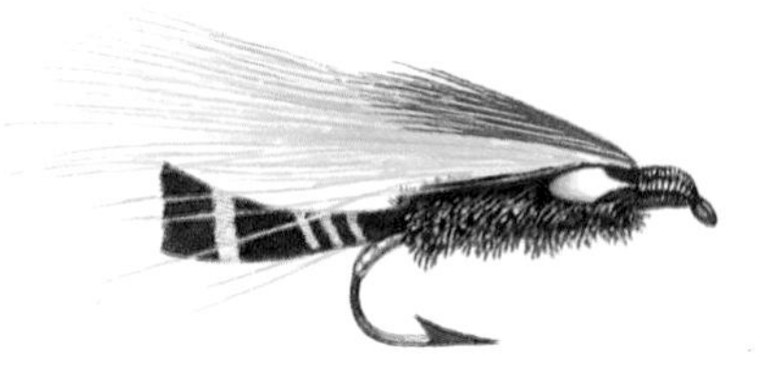

This American pattern goes back to 1929 and was first tied by William R. Edson of Portland, Maine. There is a companion fly, the Edson Dark Tiger, which has a yellow chenille body and a darker wing.

Hook Long shank 4-12.
Thread Black.
Tail Barred wood duck flank.
Body A tip of gold tinsel, followed by bronze peacock herl.
Rib None.
Hackle None.
Wing Yellow bucktail with a short tip of scarlet hackle on the top. Short jungle cock cheeks flanking.

Woolly Worm

This palmered lure is a universal favourite for many species of trout, as well as for black bass. It can come in a wide variety of body colours: green, olive, yellow, orange, brown and black. Generally speaking, all are palmered with a long-fibred grizzle hackle, though some patterns do have other coloured hackles. The Woolly Worms differ from other palmered flies in that the hackle slopes towards the eye, not towards the tail.

Hook Long shank 6-10.
Thread Black.
Tail Red wool or hackle fibres.
Body Chenille (black, yellow, green and red most popular).
Rib None.
Hackle Palmered grizzle cock sloping towards the eye.
Wing None.

Silver Darter

A classic American streamer fly created by Lew Oatman of Shushan, New York State. The Silver Darter is just one of many streamers devised by Lew Oatman to imitate a wide number of baitfish. His other flies include the Shushan Postmaster, Cut Lips and Yellow Perch.

Hook Long shank 4-10.
Thread Black.
Tail Thin slip of silver pheasant wing quill.
Body White floss.
Rib Narrow flat tinsel.
Hackle Two or three peacock sword feather points.
Wing Silver badge hackles with Jungle cock cheek optional.

Matuka Sculpin

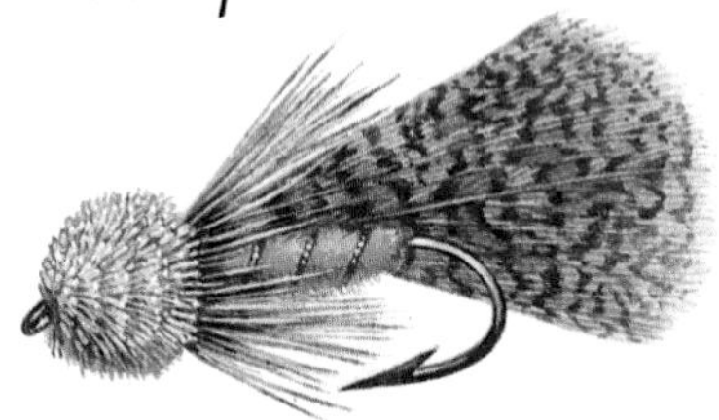

This American fly is a development of the renowned Muddler Minnow. The addition of the Matuka-style wing gives us a more realistic imitation of the Sculpin Minnow. In Great Britain the fly imitates the small fish called the Miller's Thumb or bullhead.

Hook Long shank 4-6.
Thread Amber.
Tail None (this is formed by the wing).
Body Light amber sparkle yarn.
Rib Fine oval gold tinsel.
Hackle Unclipped deer hair.
Wing Speckled hen saddle hackle, or mottled partridge tail tied matuka style. Side fins are two mottled hen body feathers on either side.
Head Clipped deer hair. The deer hair can be left natural or marked along the top with a brown felt-tipped pen.

Babine Special

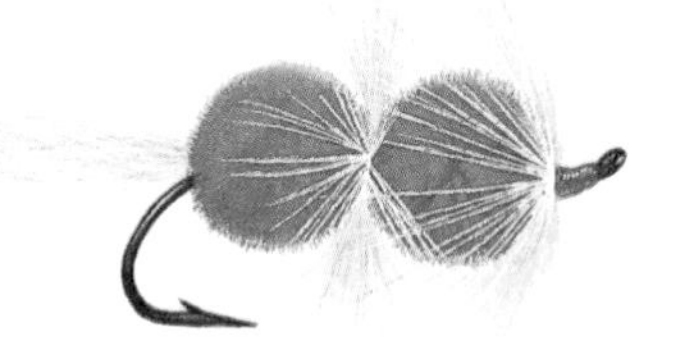

A very popular fly from the Pacific North West, this egg imitation is used right up to Alaska for many species of game fish. It is often weighted and used during the salmon spawning season. It is sometimes tied using red or pink chenille.

Hook Long shank 8-6 or salmon hook to 1/0.
Thread Fluorescent orange.
Tail White polar bear hair or calf tail.
Body In two sections: fluorescent orange chenille separated by a white cock hackle.
Rib None.
Hackle White cock hackle.
Wing None.

Polar Shrimp

This is a favourite fly for steelhead and various Pacific salmon. In smaller sizes it is often used for rainbow trout. It is used in both the western United States and Canada.

Hook Long shank 2-6 or salmon hook to 1/0.
Thread Black.
Tail Orange or red hackle tips.
Body Fluorescent orange chenille.
Rib None.
Hackle Orange cock hackle.
Wing White polar bear or calf tail.

Yellow Matuka

The Matuka style of winging comes from New Zealand, where it was used on flies for the famous lakes, Taupo and Rotorua. The Matuka is a bittern, now a protected bird so its feathers are no longer used in fly dressing. Hen pheasant flank makes a good substitute. Any fly which has the wing tied down by the ribbing tinsel is termed 'Matuka' style. This pattern, the Yellow Matuka, is a useful fly in cloudy water conditions.

Hook Standard size 4-10 or long shank 6-10.
Thread Black.
Tail None.
Body Yellow chenille.
Rib Silver or gold oval tinsel.
Hackle None.
Wing Hen pheasant flank feathers tied down with the ribbing tinsel.

Red Split Partridge

A popular New Zealand fly, again tied in the Matuka style but this time the wing is a mottled centre tail from a grey partridge. The feather is split down the middle into two halves, placed back to back and tied on top of the hook with the oval tinsel rib. Apart from the red version, other colours are sometimes used.

Hook 2-8.
Thread Black.
Tail Brown hackle fibres.
Body Red chenille.
Rib Oval silver tinsel.
Hackle Brown cock hackle fibres.
Wing Partridge tail.

Red Setter

This New Zealand fly is used by many fly fishermen in South Africa. It is tied in the same style as the Fuzzy Wuzzy, also given in this book.

Hook 2-8.
Thread Black.
Tail Brown squirrel tail.
Body Orange chenille.
Rib None.
Hackle Two: ginger cock. One hackle at the head; the other half way down the shank.

Parsons Glory

One of the classic New Zealand patterns devised by Phil Parsons, one of the pioneers of Taupo fishing. The pattern is supposed to resemble a fingerling trout.

Hook 2-10.
Thread Black or yellow.
Tail Red or orange cock fibres.
Body Yellow chenille.
Rib Oval silver tinsel.
Hackle Honey grizzle (light cree).
Wing Honey grizzle cock hackles tied matuka style. Jungle cock cheek optional.

Fuzzy Wuzzy

This fly should be fished in short sharp jerks, as it is supposed to represent the indigenous crayfish of New Zealand. With each jerk the hackles open and close in imitation of a crayfish's legs when it is swimming. The two main colours for the Fuzzy Wuzzy are black or red.

Hook 2-8.
Thread Black.
Tail Black squirrel.
Body Chenille (black, red, orange or yellow).
Rib None.
Hackle Black: one at the head; the other half way down the shank.
Wing None.

Mrs Simpson

A famous New Zealand fly pattern presumably named after Mrs Simpson, who became the Duchess of Windsor. The fly is used in the United Kingdom and also South Africa.

Hook 2-8.
Thread Black.
Tail Black squirrel tail.
Body Yellow or red wool or chenille.
Rib None.
Hackle None.
Wing Tied along the sides of the fly, cock pheasant green rump.

Yellow Rabbit

The first time I used a Rabbit fly, I was looking down onto a clear pool. I carelessly dropped my fly over the bridge to watch its action in the water. As I watched two rainbow trout give chase along with a darting perch and finally a jack pike. As with many of the New Zealand flies, a range of different coloured bodies are possible.

Hook 2-8.
Thread Black or yellow.
Tail Cock hackle fibres, red or yellow.
Body Yellow chenille (other colours can be used).
Rib Oval silver tinsel.
Hackle Yellow cock hackle (other colours to match body colour).
Wing A strip of rabbit skin.

Walker's Killer

The premier fly of the Republic of South Africa. It appears that any South African fly fisherman who did not use this pattern would be considered a heretic. In recent years the fly has received a degree of popularity in the UK, due mainly to the fact that a high proportion of the commercially-dressed flies sold in this country are tied in South and East Africa.

Hook Long shank 4-10 or normal shank 2-8.
Thread Black.
Tail Black squirrel.
Body Red chenille.
Rib None.
Hackle None.
Wing Striped partridge hackle, tied along the sides in three sets of three hackles.

Vlei Kurper

A lure tried to represent one of the small baitfish of South Africa, the vlei kurper. Michael G. Saloman in his book *Freshwater Fishing in South Africa* strongly recommends this fly.

Hook Long shank 6.
Thread Black.
Tail None.
Body Yellow floss with final quarter fluorescent red wool.
Rib Silver wire.
Hackle Yellow bucktail.
Wing Dark green bucktail over medium green bucktail. Copper lurex either side as cheeks.
Head Black with white painted eye.

Wiggle Sculpin

Two Sculpin imitations, the Muddler and the Matuka Sculpin, have already been mentioned. This fly from Roman Moser of Austria is even more realistic. It moves exactly like the small fish it is supposed to represent. Roman Moser's flies are strongly influenced by such American tyers as Dave Whitlock.

Hook Lowater salmon size 6 linked during tying to a similar size hook with bend removed.
Thread Brown.
Tail Polypropylene fibres.
Body Extra thick chenille, marked after tying with a brown waterproof pen.
Rib None.
Hackle None, but two tufts of polypropylene as side gills.
Wing None.
Head Two yellow glass eyes.

Seatrout flies

The mercurial seatrout probably captures the imagination of a greater number of anglers than even the more predictable salmon. Often described as fickle, this shy, migratory version of the brown trout is seldom (as a general rule) taken during the hours of daylight, and those anglers who seek the silver fish share their fishing hours with bats and other creatures of the night.

The seatrout will run up rivers even in low-water conditions. In fact the saying goes: 'Seatrout will run on a fall of dew'. Salmon, on the other hand, require the prompting of a good spate to make the journey from the sea to their spawning grounds.

There are two types of seatrout: the large, solitary fish that usually make their journey early in the year; and the much smaller school seatrout of the summer months, which sometimes run up the river in large shoals.

Seatrout have always been somewhat of an enigma. Until recent times they were thought to be a distinct species, and had the scientific name *Salmo trutta*. The brown trout was known as *Salmo fario*. Scientists have since proved that the two fish are one and the same, and the term *Salmo trutta* now embraces both. The question of what prompts the brown trout of one river to migrate to the sea, while trout in adjacent waters remain in the river, is part of the mystery surrounding the fish. And it is not entirely answered by the food supply in the rivers.

Just like the salmon, the seatrout is reputed not to feed in freshwater. It is supposed that its return from the sea is strongly motivated by the instinct to spawn, which overcomes any desire to feed. Autopsies on the stomach contents of caught fish reveal very little in the way of food; just the odd insect, but that is all. There is no evidence of strong feeding like one would find in a brown or rainbow trout, which, of course, endorses the non-feeding hypothesis. However, quite a large proportion of school seatrout do not spawn on their first visit back to freshwater, and it is believed that such fish will in fact feed.

Seatrout can be caught on most types of fly. Flies given in the dry fly and wet fly sections, the streamers and bucktails, and also the salmon flies, will all take seatrout. In the past, all seatrout flies were either scaled-down salmon flies or scaled-up trout flies, and to a certain extent this holds true today. However, modern flies by Hugh Falkus are devised solely for seatrout. They have been created to catch fish in their various taking moods.

The selection of patterns given in this section covers most of the different types of fly or lure required to take the seatrout: larger versions of standard wet flies; floating, wake-forming surface lures; sub-surface lures; and dapping flies.

On some nights the seatrout is content to take small, sub-surface flies; on other occasions a much larger lure is required to prompt any sort of response from the fish. At other times again, with a sudden awakening from its shy daylight torpor, the seatrout will explode with a burst of energy, wanting nothing but a large fly waked across the surface. Local knowledge of rivers and flies is a big asset when seatrout fishing, otherwise a system of trial and error is needed to select the right pattern.

Fickle and finnicky, shy and secretive, may well sum up the seatrout, but if the salmon is the King of fish, then surely the seatrout must be the Prince of the river.

Medicine Fly

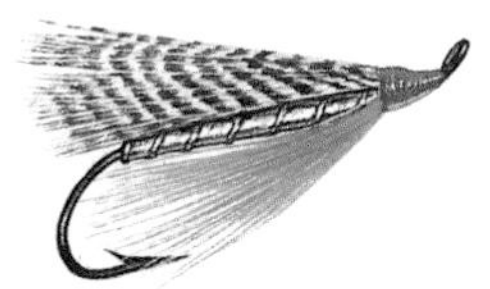

Hugh Falkus's version of the Teal Blue and Silver. Falkus is probably Britain's leading authority on the seatrout. His book *Seatrout Fishing* (Witherby) is considered to be the best work on the subject. This fly must be tied slim for best effect.

Hook Lowater salmon hook sizes 2, 4, 6.
Thread Black or red.
Tail None.
Body Flat silver tinsel.
Rib Oval silver tinsel (optional).
Hackle Light blue.
Wing Teal, mallard or widgeon.

Goat's Toe

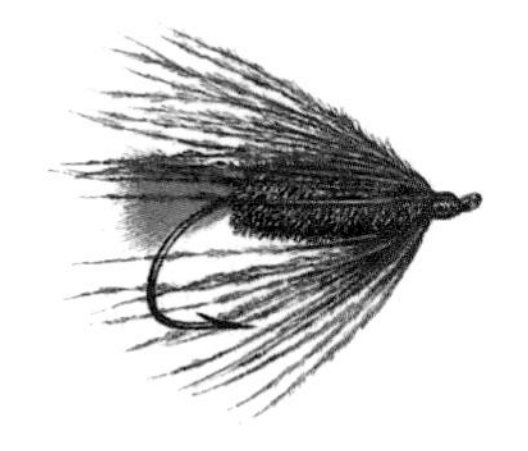

An unusual pattern favoured by many who fish in the north of Scotland and the Islands. This fly is also sometimes used for salmon.

Hook 10-6.
Thread Black.
Tail Scarlet wool.
Body Bronze peacock herl.
Rib Scarlet floss.
Hackle Blue peacock neck feather.
Wing None.

Secret Weapon

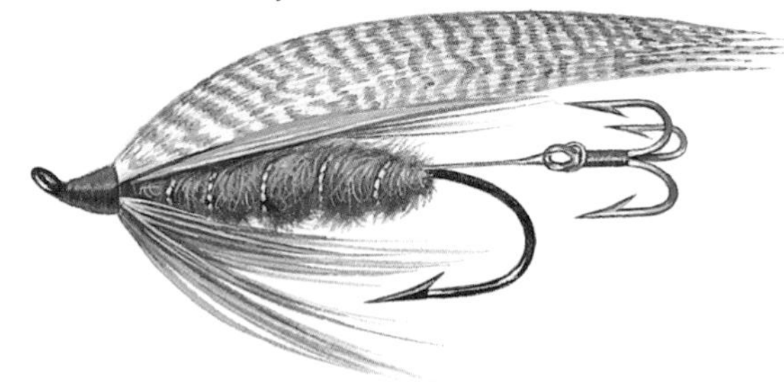

Another of the Falkus seatrout patterns, this one has a flying treble projecting from the tail. The fly is sometimes used with a few maggots attached to the main hook. The treble seems to be the answer for the fish that come short to the fly or maggot-laden fly.

Hook Front hook lowater salmon 4-6, or standard size 4-6, plus linked small flying treble 14-16.
Thread Red.
Tail None.
Body Light brown fur dubbing.
Rib Gold wire (optional).
Hackle Natural red hen.
Wing Bronze mallard.
Head Red.

Teal Blue and Silver

This is a traditional pattern for seatrout. It is one of a series of Teal flies used for both lake and migratory trout. Seatrout when newly arrived from the sea are often attracted by the colour blue.

Hook 10-6.
Thread Black.
Tail Golden pheasant tippet.
Body Flat silver tinsel.
Rib Oval silver tinsel.
Hackle Light blue.
Wing Teal flank.

William's Favourite

This is a favourite pattern of mine; one that I used on the rivers of my boyhood in North Wales. The fly was devised by the father of Courtney Williams, author of *The Dictionary of Trout Flies*. It is very similar to the Black Pennell, the only difference being in the tail; the Pennell uses golden pheasant for the tail.

Hook 6-12.
Thread Black.
Tail Black cock hackle fibres.
Body Black silk.
Rib Silver tinsel.
Hackle Black hen or soft-fibred cock.
Wing None.

Dai Ben

One of the best-known sewin flies from the principality of Wales, the Dai Ben is a fly from the River Towy. Moc Morgan, the authority on flies and all things Welsh, states that it was named by a well-known Towy angler, David Benjamin Glyn Davies of Abergwili, Dai Ben being the diminutive of David Benjamin. I have seen this fly dressed with Andalusian blue hackles, a breed of cockerel kept by many keen anglers/fly-dressers in South and Mid Wales.

Hook 6-10.
Thread Black.
Tail Honey dun fibres.
Body Rabbit fur.
Rib Flat silver tinsel.
Hackle Honey dun or blue dun.
Wing None.

Conway Red

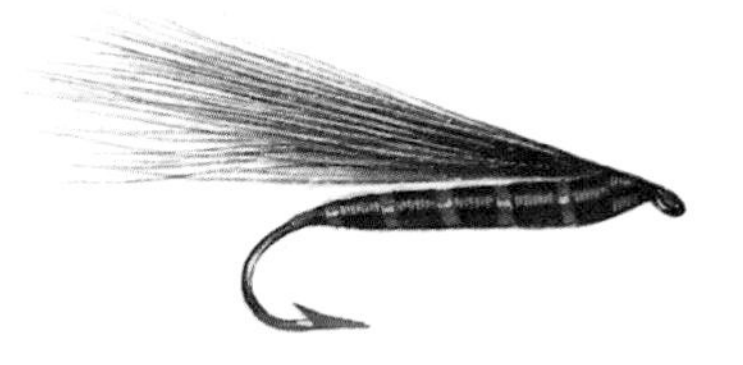

A North Wales pattern hailing from the River Conway area. The badger is now a protected mammal and any hair available usually comes from animals killed on the roads. Grey squirrel works just as well for the wing.

Hook 6-8.
Thread Black.
Tail Red silk (optional).
Body Black floss.
Rib Red tinsel.
Hackle None.
Wing Badger.

Harry Tom

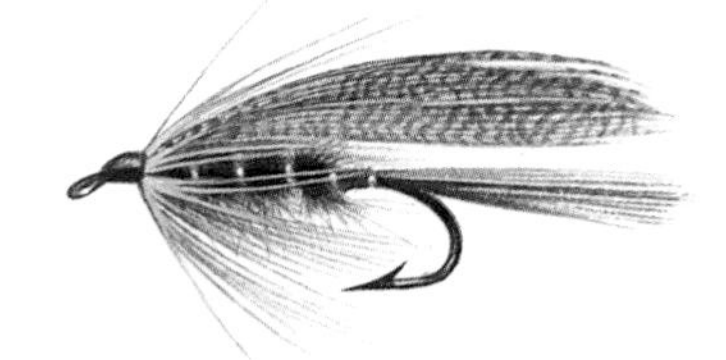

A fly from North Wales, hailing from the Ogwen valley area. It is a brother fly of the Dai Ben, and is often used as a dropper fly.

Hook 8-10.
Thread Black or brown.
Tail Honey dun hackle fibres.
Body Rabbit fur.
Rib Silver wire.
Hackle Honey dun.
Wing Bronze mallard.

Ke He

A popular hackled pattern used in the north of Scotland. It is a slight variation on the Red Tag. The fly goes back to the early 1930s and was invented by two gentlemen by the names of Heddle and Kemp, the first few letters of their names providing the name of the fly.

Hook 6-10.
Thread Black.
Tail Red wool and golden pheasant tippets.
Body Bronze peacock herl.
Rib None.
Hackle Natural red hen.
Wing None.

Kingsmill

This pattern was created by T. C. Kingsmill Moore, author of the fishing classic *A Man May Fish*. This fly is equally good for lake trout as for seatrout. Kingsmill Moore's flies have revived in popularity in the last few years due to the re-publication of his book.

Hook 6-8.
Thread Black.
Tail Golden pheasant tippet.
Body Black ostrich herl.
Rib Silver wire.
Hackle Black cock.
Wing Rook wing quill; jungle cock cheeks with golden pheasant crest over.

Surface Muddler

Quite often, seatrout are attracted to a fly waked in the surface, rather than a sunk pattern. This Muddler-style fly makes an ideal lure for such fishing. It can be tied as a single or as a tandem fly. The example shown uses blue-dyed deer hair although any colour seems to work; white, black and natural being very effective.

Hook 4-8 (or long shank 8-6).
Thread Black or blue.
Tail None.
Body Oval silver tinsel.
Rib None.
Hackle Unclipped deer hair.
Wing Blue bucktail or other blue hair.
Head Blue clipped deer hair (undyed deer hair also works).

Falkus Sunk Lure

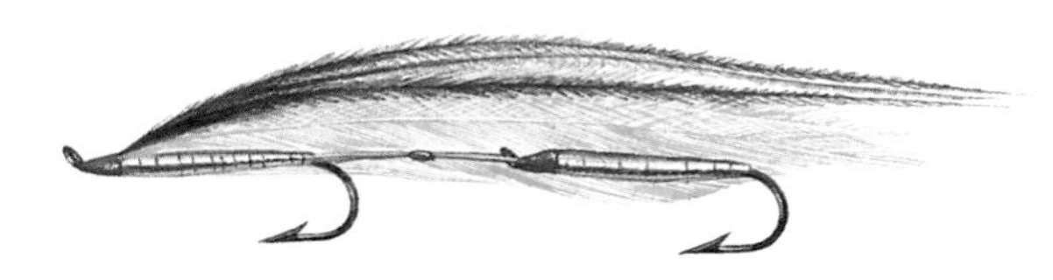

A tandem fly devised by Hugh Falkus. This is more or less an elongated version of his Medicine Fly.

Hook Two size 8 in tandem.
Thread Red.
Tail None.
Body Flat silver tinsel.
Rib Oval silver tinsel (optional).
Hackle Blue (optional).
Wing Four light blue cock hackles, with peacock herl strands over.

Huw Nain

The slightly degoratory English expression 'Mummy's boy' has a close Welsh equivalent in this fly. 'Huw' stands for Hugh, and 'Nain' means grandmother. Moc Morgan suggests that the name could come from a person who was raised by his grandmother. There are definitely stranger nicknames and nickname-combinations in Wales. The fly is a pattern from the Conway area, from Dolwyddelan, where a ruined Welsh castle overlooks the river.

Hook 6-8.
Thread Black.
Tail Golden pheasant tippet.
Body Rear half golden olive, front half grey seal's fur.
Rib Silver wire.
Hackle Brown partridge.
Wing Hen pheasant wing quill.

Silver March Brown

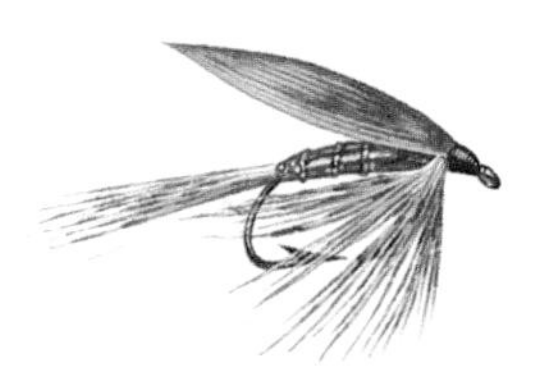

There are many March Brown patterns. This flashy, silver fly is an attractor version of the fly and is used by stillwater, river and seatrout fisherman alike. It is a traditional fly that has stood the test of time.

Hook 6-10.
Thread Black or brown.
Tail Partridge hackle fibres.
Body Flat silver tinsel.
Rib Oval silver tinsel.
Hackle Brown partridge.
Wing Hen pheasant wing quill.

Wormfly

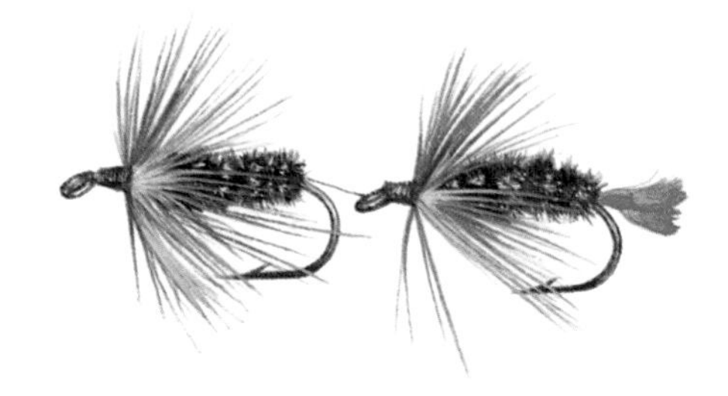

This fly could just as well appear in the section on lures. It is equally at home in both camps, and is equally attractive to stillwater trout and seatrout.

Hook Two size 8 in tandem.
Thread Black.
Tail Red wool (sometimes both hooks are tailed).
Body Peacock herl.
Rib None.
Hackle Natural red hen.
Wing None.

Silver Invicta

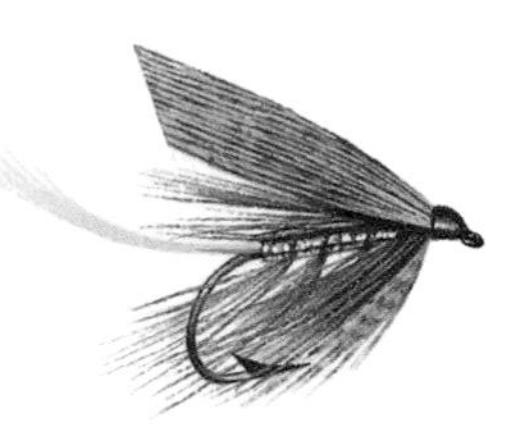

In Wales, this fly is sometimes called the Silver Knicker. It is very similar to the Silver March Brown, and is a good alternative to it. This fly is a brighter version of the more sober Invicta, which is also very effective for seatrout.

Hook 6-8.
Thread Black.
Tail Golden pheasant crest.
Body Flat silver tinsel.
Rib Oval silver tinsel.
Hackle Palmered light natural red cock; blue jay at the throat.
Wing Hen pheasant centre tail (usually tied using wing quill as it is easier).

Connemara Black

A classic favourite fly from Ireland, this pattern is used for both lake trout and seatrout. It is popular in South Africa as a standard trout pattern.

Hook 6-10.
Thread Black.
Tail Golden pheasant crest.
Body Black seal's fur.
Rib Fine oval silver tinsel.
Hackle Black with blue jay at the throat.
Wing Bronze mallard.

Donegal Blue

With a name like this, it just has to be Irish. Some versions are tied with a chenille body, and sometimes the rib is omitted.

Hook 8-12.
Thread Black.
Tail None.
Body Dark blue seal's fur.
Rib Flat silver tinsel (sometimes embossed).
Hackle Long-fibred cock hackle.
Wing None.

Pheasant and Yellow

An Irish lake pattern used in some areas for seatrout. Dyed-red swan or goose can be used as a substitute for the scarcer red macaw.

Hook 6-10.
Thread Black or yellow.
Tail Either red macaw or golden pheasant tippet.
Body Yellow seal's fur.
Rib Oval gold tinsel.
Hackle Blue jay.
Wing Hen pheasant wing quill.

Donegal Olive

This pattern hails from the North West of Ireland. There is no connection between it and the Donegal Blue, in design or conception. An interesting feature of this fly is the combination of different colours required to make up the overall body hue.

Hook 8-12.
Thread Black.
Tail Grey mallard flank fibres.
Body Mixed seal's fur in equal parts: green, yellow, scarlet, light blue, golden olive and orange.
Rib Oval gold.
Hackle Dark claret.
Wing Bronze mallard.

Alexandra Gold

A gold-bodied version of the traditional Alexandra fly, which has a silver body. Both are used for trout and seatrout. Another version has cheeks of Jungle cock feathers rather than red ibis.

Hook 6-10.
Thread Black.
Tail Red ibis substitute, plus peacock sword.
Body Gold flat tinsel.
Rib Oval gold tinsel.
Hackle Black.
Wing Green peacock sword flanked by red ibis.

Mallard and Blue

One of the Mallard series of flies, all of which are used for seatrout and trout. Other colours include Mallard and Claret, Mallard and Green, Mallard and Yellow, Mallard and Black, Mallard and Silver, Mallard and Gold.

Hook 6-12.
Thread Black.
Tail Golden pheasant tippet.
Body Blue seal's fur.
Rib Oval silver or gold tinsel.
Hackle Natural red hen or blue-dyed hen.
Wing Bronze mallard.

Elver Lure

A silver adaptation of Arthur Ransome's Elver Lure, this fly, tied on inch tubes, is a highly successful fly in the West Country of England.

Hook Lowater salmon 4-8.
Thread Black or red.
Tail None.
Body Flat silver tinsel.
Rib Oval silver tinsel.
Hackle Light blue guinea fowl or light blue hen.
Wing Vulturine galena with Jungle cock either side.

Teal and Green

Another of the Teal series. Like all the other colours it can be used for trout as well as seatrout.

Hook 6-10.
Thread Black or green.
Tail Golden pheasant tippet.
Body Green seal's fur.
Rib Oval gold or silver tinsel.
Hackle Green or black.
Wing Teal flank.

Haslam

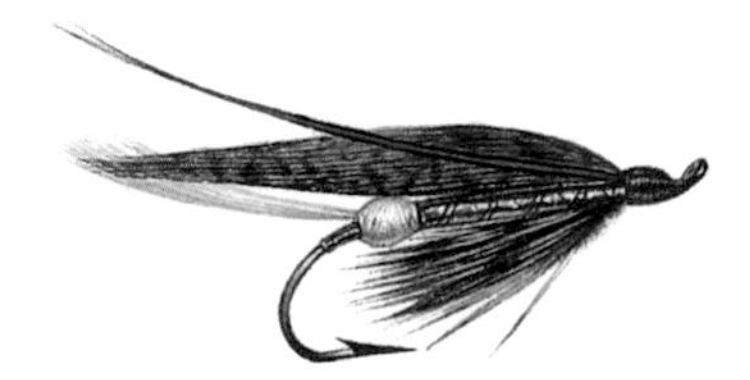

This salmon fly from the Dovey area of Mid Wales is often used in smaller sizes for seatrout on both the River Dovey, and the River Mawddach further north.

Hook 6-8.
Thread Black.
Tail Golden pheasant crest.
Body Tag-flat silver tinsel; butt-white wool; body-flat silver tinsel.
Rib Oval silver tinsel.
Hackle White or badger palmered (optional), with blue jay or guinea fowl at the throat.
Wing Hen pheasant tail.
Horns Blue macaw.

The Kudling

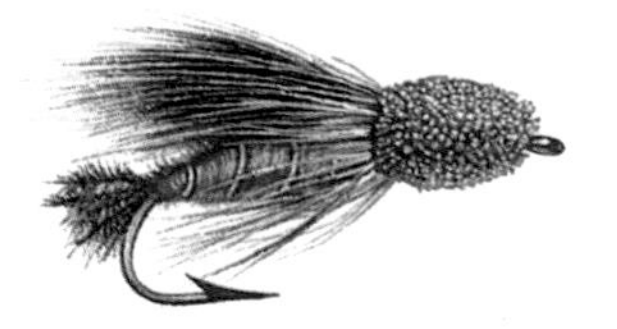

A Danish Muddler-style fly devised by Tom Petersen. This fly is used also for trout. Some Danish fly fishermen use this fly as a saltwater lure. The Kudling works well in rivers and streams with a population of sticklebacks.

Hook Long shank 6-8.
Thread Black.
Tail Bronze peacock herl.
Body Grey wool, with bronze peacock herl over the back.
Rib Fine silver oval.
Hackle Unclipped deer hair.
Wing Grey squirrel.
Head Clipped deer hair muddler head.

The Mørum Fly

A fly devised by Jan Grünwald for seatrout and salmon in Swedish waters, in particular the River Mørum. This fly and the Black Shank work well in highly-coloured waters.

Hook Double salmon size 4-10.
Thread Black.
Tail None. Tag oval silver followed by red floss.
Body Black floss.
Rib Oval silver tinsel.
Hackle Palmered red golden pheasant body feather.
Wing Teal at the throat; golden pheasant tippet, with golden pheasant tail over.

Black Shank

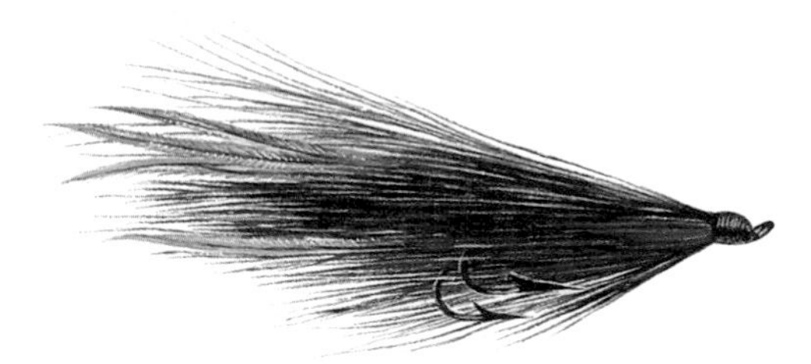

Another fly for the River Mørum in Sweden. As you will see, this fly has no body dressing, but it has plenty of action in the long-flowing hair wing.

Hook Double salmon 4-10.
Thread Black.
Tail None.
Body None.
Rib None.
Hackle Black squirrel.
Wing Red-dyed squirrel with black squirrel over, along with a few strands of bronze peacock herl tied fairly long.

Badger Seatrout Streamer

A Danish seatrout streamer fly in the typical streamer style of tying. It imitates a small fish. Many of the streamer patterns given in this book work extremely well for the migratory trout.

Hook Long shank 6-8.
Thread Black.
Tail None.
Body Flat silver tinsel.
Rib Oval silver tinsel (optional).
Hackle Collar hackle, badger.
Wing None.

Krogsgaard No. 1

This is one of a series of nine flies from Denmark which date back to the 1920s. They were devised by Olav Krogsgard and J. Kr. Findal, and must be considered as standard patterns for Scandinavia. They are used for both brown trout and seatrout. All of them are tied in the caddis style of dressing.

Hook 4-8.
Thread Black.
Tail None.
Body Olive floss.
Rib Oval gold (optional).
Hackle Natural red hen.
Wing Two strips of brown hen wing quill, with two strips of white duck on the inside.

Krogsgaard No. 2

This brown fly is second in the series and possibly one of the best for seatrout.

Hook 4-8.
Thread Brown.
Tail None.
Body Brown floss.
Rib Oval gold.
Hackle Ginger.
Wing Two strips of cinnamon turkey, with two strips of white duck on the inside.

Telemarkskongen

A Norwegian fly which is very similar in colouration to the standard Teal Blue and Silver. The main difference lies in the tail. In this pattern the tail is a bright, attractive red.

Hook 6-8.
Thread Black.
Tail Red ibis substitute.
Body Flat silver tinsel.
Rib Oval silver tinsel.
Hackle Light blue.
Wing Grey mallard flank.

Trogersen's Fancy

A bright seatrout pattern from Norway. It is easily identified by its unusually bright red wing.

Hook 6-10.
Thread Black.
Tail Golden pheasant tippet.
Body Black floss.
Rib Silver wire or oval.
Hackle Black hen.
Wing Red-dyed swan or goose.

Dapping Flies

On many Scottish and Irish lakes that are the headwaters of several rivers, it is possible to take seatrout during the hours of daylight by fishing with dapping flies. It is difficult to say what these flies represent. What is certain is that the seatrout find them extremely attractive. Colour does not appear to matter, though black and blue seem to be most popular. Large, heavily-hackled versions of some of the standard flies are also used for such fishing, such as the Pennells and Palmer flies. Realistic, insect-imitating lures, such as the Cranefly, are also used.

Hook Lowater salmon or Wilson Dry Salmon 4-8.
Thread Black.
Tail Black cock hackle fibres (optional).
Body None as such; hackles form the body.
Rib None.
Hackle Heavily palmered black cock with blue at the head.
Wing Black squirrel (optional).

Grayling flies

This section of the book is devoted to a small number of grayling flies, though it must be said that flies which take a trout will, more often than not, also take grayling.

The grayling is truly a game fish, although it is considered by some to be a coarse fish and a pest of the worst kind on a trout stream. They believe that the grayling competes with the trout for food and for comfortable lies in the stream, to the ruination of the trout fishing. If one happened to be carefully stalking a trout of a good size, and one's well-presented dry fly was suddenly seized by a small grayling, it would be very annoying. However, the 'Grey Lady of the Stream' is now appreciated for what she is, a game fish with all the true sporting attributes: a riser to the dry fly; and a fish that, when hooked, offers a good fight, perhaps not quite as spectacular as a trout, but nevertheless a fish in top condition can put up a dour struggle.

The grayling (*Thymallus thymallus*) was at one time placed firmly in the family of the salmon (*Salmo thymallus*), the common factor between the grayling and members of the salmon family being the tiny fin to the rear of the back, the adipose fin. However, the grayling is now placed in a family of its own. The shape of its mouth indicates that it is primarily a bottom-feeder. Its top lip overhangs the bottom lip, and the fish grubs around the bottom of the river, seeking out such creatures as caddis larvae, mayfly nymphs, water lice, freshwater shrimps, snails and any other wriggling nymphs or pupae that it comes across. With all the subaquatic food available to it, it is something of a mystery why such a bottom-feeding fish will rise to take a dry fly off the surface.

Unlike the brown trout, the grayling spawns in the spring and early summer. So when the normal trout season ends, the Grey Lady is at her peak of fitness and provides a worthy challenge to the fly-fisherman, and a winter dimension to fly-fishing. There is nothing better on a crisp autumn or winter morning, when your breath is caught in a mist and when your feet crunch through the frost-covered grass, to approach a river and see the dimpled rises of feeding grayling. Such mornings are a blessing and should be cherished. To my mind, the presence of grayling enhances the value of any trout stream.

Many of the fancy flies tied specifically for this fish possess small, bulky peacock herl bodies and, inevitably, a small red tail. This formula was probably considered to be the most killing by early fly-dressers, but I must admit to being content to use more sober, natural-looking flies, with an occasional flirtation with the Sturdy's Fancy. Concerning the size of hook to use, I fish dry on a size that is seldom larger than 16; if I am fishing a wet pattern or a nymph, I move up to a size 12.

Further information on grayling fishing can be found in T. E. Pritt's *Book of the Grayling* published in 1888; *Grayling Fishing* by W. Carter Platts, published just before World War II; *Grayling in South Country Streams* by H. A. Rolt; and finally, *The Grayling Angler* by John Roberts, published in 1982.

Sturdy's Fancy

A favourite fly for the rivers of the Yorkshire Dales, UK. It has been referred to as 'an old man's fly' because its white hackle can be easily seen on the surface of the water. I am sure it is not just an old man's fly, but I must admit that this pattern had a higher degree of clarity than some other flies I had been using.

Hook 14-16.
Thread Black.
Tail Bright red wool or silk.
Body Green peacock herl.
Rib None.
Hackle White cock.
Wing None.

Treacle Parkin

Sometimes called the Orange Tag, for obvious reasons. Some anglers of the Hampshire chalk-streams swear by this fly for trout as well as grayling. The French have a similar pattern with a white wool tail which they use for both species.

Hook 12-16.
Thread Black.
Tail Orange wool.
Body Bronze peacock herl.
Rib None.
Hackle Natural red cock.
Wing None.

Red Tag

One of the standard old-time flies for trout and grayling. According to Courtney Williams, the fly originated in Worcestershire. It is now firmly established all over the world.

Hook 12-16.
Thread Black.
Tail Red wool.
Body Bronze peacock herl.
Rib None.
Hackle Natural red cock.
Wing None.

Green Insect

One of the classic grayling patterns, used wherever the grayling is fished, from Yorkshire to Hampshire. Some anglers deem it to be an imitation of a small caterpillar or similar grub. It is sometimes, like many grayling flies, given a red attractor tail.

Hook 14-16.
Thread Black or green.
Tail None.
Body Green peacock herl.
Rib None.
Hackle Grey hen.
Wing None.

Sanctuary

A very good, natural-looking artificial fly that works well wet or dry, for both trout and grayling.

Hook 14-16.
Thread Black or brown.
Tail None.
Body Dark hare's ear.
Rib Flat gold tinsel.
Hackle Coch-y-bonddu.
Wing None.

Grayling Witch

There are a number of Witch patterns: Gold Witch, Silver Witch, and The Witch. They were originally the creation of H. A. Rolt. The pattern given here was the creation of Roger Wooley. The Witch flies date from the beginning of the twentieth century.

Hook 14-16.
Thread Black.
Tail Red floss.
Body Green peacock herl.
Rib Flat silver.
Hackle Pale blue dun.
Wing None.

Brooke's Fancy

Another grayling fly that does not have a red tail. Like most specific grayling flies, it can be classed only as an attractor fly, rather than an imitation of a specific insect.

Hook 12-14.
Thread Black.
Tail None.
Body Purple silk.
Rib Peacock herl.
Hackle White.
Wing None.

Orange Otter

The original dressing for this colourful fly utilized dyed otter fur as the body medium. Now that the otter is a protected animal, dyed seal's fur is usually used instead. However, seals are also an emotive subject now, so most modern fly-dressers use various man-made fibres.

Hook 14-16.
Thread Orange.
Tail Natural red cock hackle fibres.
Body Orange seal's fur, or substitute. The original used dyed otter fur.
Rib None.
Hackle Natural red tied in the middle of the hook.
Wing None.

Kill Devil Spider

This is a particular favourite of mine. I have lost count of the number of decent trout I have caught with this grayling fly. I fish it, for both species, in the surface film where the fish probably mistake it for some swamped terrestrial. I have used it down to size 20. There are a number of Kill Devil Spiders, Black (depicted here) and Brown being the most popular. They are attributed to David Foster of Ashbourne, Derbyshire.

Hook 16-18.
Thread Black.
Tail None.
Body Bronze peacock herl.
Rib None.
Hackle Long-fibre black cock or hen.
Wing None.

Grayling Lure

Frank Sawyer's answer to the grayling. This no-hackle, no-wing fly is very reminiscent of the grayling grubs of Francis Francis. The wool for Sawyer's Grayling Lure was the proprietary brand Chadwicks 477. The other name for this fly is Killer Bug.

Hook 12-14.
Thread Brown.
Tail None.
Body Fawn wool with pinkish tone when wet, over copper wire.
Rib None.
Hackle None.
Wing None.

Tommy's Favourite

This fly, dating back to the nineteenth century, is now somewhat out of favour, mainly due to the lack of yellow and blue macaw tail feathers. If they are obtainable, they can be very expensive. A strand of yellow and a strand of blue goose or swan fibre can be twisted together to achieve the same effect.

Hook 14-16.
Thread Black.
Tail Red wool.
Body Yellow/blue macaw tail fibres.
Rib None.
Hackle Blue dun.
Wing None.

Red Badger

One of a series of Badger flies still in use today. They include the Blue Badger, Black Badger, Silver Badger, Gold Badger and the one depicted here, the Red Badger.

Hook 14-16.
Thread Black.
Tail None.
Body Red floss.
Rib Silver wire.
Hackle Palmered badger cock.
Wing None.

Salmon flies

No fish has captured the imagination more than the King of Fish, the Atlantic salmon. Gone are the days when apprentices complained to their masters at having to eat salmon every day. Now the salmon has a high commercial value because of its scarcity. Commercial netting, disease, pollution and, on some rivers, water abstraction, have all contributed to the great drop in the number of salmon that return to the rivers. The value of fish that do come back has in itself created a new industry based on the highly organized, ruthless poaching gangs, who can earn themselves thousands of pounds for a night's work. Taking all these factors into account, salmon fishing is not what it used to be. Anyone who catches a salmon today can consider himself lucky, and big catches to the rod are a rarity.

After hatching, the salmon spends the first part of its life in the river. At this stage it resembles a small trout and behaves in much the same way, until it reaches a size of about 5½-6 in (14-15.25 cm). It then undergoes a distinct physical change, triggered by a release of hormones from the pituitary and thyroid glands. It develops a silvery appearance, and begins its migration to the sea. The young salmon, now called a smolt, migrates up into the north Atlantic, where the rich feeding zones are found. In due course it will return to the river of its birth, driven on by the powerful desire to spawn. Fish that return after only one year at sea are called grilse.

It is accepted that salmon do not feed once they have returned to freshwater, their desire to travel up the rivers to spawn is so great. A similar hormone change is probably also responsible for this. They rely completely on the fat they have put on during their sojourn in the sea. All sorts of theories are, therefore, put forward to explain why a salmon takes an artificial fly: from curiosity, from anger, or perhaps from a lapse in its concentration on spawning. No one can really tell. We just give thanks for the fact that it will, from time to time, seize our flies.

Glancing at the pages of some of the books on salmon fishing from yesteryear, the reader cannot fail to be impressed by plates of the most ornate and decorative flies, using all manner of rare and exotic feathers. Today's angler tends to use less complex flies. Many of the ingredients used in those Victorian creations are not available to the modern fly-dresser. One can only substitute with alternative materials to a certain extent without changing the appearance of the fly completely, thereby losing the purity of the dressing. In addition, the cost of some feathers that are still available can be quite prohibitive, and old stocks of materials do not last forever. The Jungle cock feathers used as a cheek on many flies are an example; the Sonnerat's Jungle fowl is now a protected species so the only feathers available come from birds specially bred for fly-tying purposes. A cape from such a bird can cost as much as £35. The angler has therefore been forced to seek less costly flies made from more readily available materials. Hence the birth of simple 'hair-wing' patterns.

Styles and types of salmon flies

Simple strip wings

These flies, as the name suggests, have a very simple wing of a single feather set back to back. They are the easiest of the salmon flies to tie. Examples given in the patterns are Skiri, Kola Fly and Sweep.

Whole feather wings

Jungle cock, golden pheasant, and peacock sword are classed as whole feathers when set back to back on top of the hook. The Em Fly is an example of this style of fly.

Mixed wing

An ornate wing made up of individual strands of various feathers married to one another to form a multi-coloured wing. An example is the Mar Lodge.

Built wing

Some fly-dressers do not make the distinction between this style and the preceeding one. The wing is slightly more complex. A fully-dressed Yellow Torrish is an example of this type of fly.

Herl wings

This style is not often used these days in salmon flies. An Alexandra (see page 50), if enlarged to salmon size, would be classed as a herl wing.

Dee strip

This type of fly, usually tied on quite large salmon irons, was fished on the Aberdeenshire Dee. The main feature of these flies was the narrow strips of turkey tail set splayed on top of the hook, and the use of long trailing heron hackles. A tube fly version of the Akroyd is given in the text.

Spey flies

As the name suggests, these flies were used on the River Spey, and like the Dee flies they relied on ultra-large hackles, sometimes heron, and had wing strips of brown mallard flank. Also like the Dee flies, they were fished in large sizes.

Shrimps and prawns

Shrimp patterns are sometimes described as 'grubs'; the Usk Grub is such a fly. The General Practitioner is perhaps the best example of a prawn pattern.

Hair-wings

Most salmon anglers today fish with hair-wing flies.

It is generally accepted that the first hair-wing salmon flies were created in North America, though it is possible that some Scottish ghillie may have tied a few hairs from his dog onto a hook and caught a fish with such a fly. Hair-wings can be tied on conventional hooks, doubles and trebles. Some North American patterns use ordinary long-shank streamer hooks. Even the most hard-bitten traditionalist has now come to accept the hair-wing fly. Apart from its obvious effectiveness in taking fish, it also makes economic sense, for these flies can be produced for a fraction of the cost of a fully-dressed salmon fly.

All traditional feather-wing flies can be converted to hair-wings.

Tube flies

Two examples of this form of fly are given in the text, the Akroyd and the Tosh. Many anglers prefer to use the tube fly, with its subtle treble, than the normal salmon hook. Most conventional patterns can easily be adapted to the tube style. There are no hard and fast rules about dressing tube flies; this is left to the individual tyer.

Waddington lures

This type of salmon fly was invented by Richard Waddington and is, in effect, very similar to the tube fly. Instead of a tube of nylon or metal, the fly is tyed on a shank onto which a treble has been slipped. Like all hair-wings and tube flies, the actual dressings are left to the tyer. Usually the bodies of the original flies are retained; the wing is created with whatever is available. The Jock Scott is an example of a Waddington dressing.

Low-water salmon flies

These are standard salmon flies, both feather and hair-wing, but are much abbreviated in the dressing for low-water summer conditions. The distinguishing features are the lightness of the hook, and the size of the dressing compared with the size of the hook. The Heggli is an example of such a fly.

Durham Ranger

One of the great nineteenth-century Tweed flies, which is still used in both feather and hair-wing versions. The fly was devised by James Wright of Sprouston Kelso, Scotland. This fly is used when a bright fly is required.

The original full dressing is given here.

Hook All sizes.
Thread Black.
Tag Round silver tinsel (sometimes also yellow floss).
Tail Topping and Indian crow.
Butt Black ostrich herl.
Body Lemon floss, orange, fiery brown, and black seal's fur.
Rib Flat silver tinsel and twist (oval silver tinsel more usual).
Hackle Palmered, dyed yellow along the body, light blue at throat.
Wing A pair of Jungle cock feathers, covered over three-quarters of length by golden pheasant tippets back to back, with topping over.
Sides Jungle cock.
Cheek Blue chatterer (substitute).
Horns Blue/yellow macaw (optional).
Head Black.

Yellow Torrish

This fly is popular in Scotland and down as far as the West Country. It is particularly effective in coloured water. The hair-wing version is a much simpler affair; a wing of yellow hair only, is often used.

Hook All sizes.
Thread Black.
Tag Round silver tinsel, yellow floss silk.
Tail Golden pheasant topping, and ibis.
Butt Black ostrich herl.
Body In two parts, silver tinsel, divided by black ostrich herl.
Rib Oval silver tinsel.
Hackle Yellow.
Wing Two strips: white-tipped turkey, bustard, peacock wing, guinea fowl, golden pheasant tail, red and blue swan, brown mallard; golden pheasant topping over all.
Cheeks Jungle cock.
Head Black.

Mar Lodge

Sir Herbert Maxwell described this fly as being very tasteful, best used in large sizes, and used on the Aberdeenshire Dee. Since his day, the fly has been used on many rivers.

The original full dressing is given here.

Hook All sizes.
Thread Black.
Tag Round silver tinsel.
Tail Golden pheasant topping and jungle cock.
Butt Black ostrich herl.
Body In three parts: first and third embossed silver tinsel; second, black silk.
Rib Oval silver tinsel.
Hackle Guinea fowl.
Wing Yellow, red, blue swan, strips of peacock wing, summer duck, grey mallard, dark mottled turkey, golden pheasant tail, peacock sword over all.
Sides Barred summer duck with golden pheasant toppings over wing.
Head Black.

Blue Doctor

One of a series of 'Doctor' flies, others being the Black Doctor and the original Silver Doctor. All have the same wing. Hair-wing versions are available for all three. In smaller sizes, with a less complicated wing, they are also used for seatrout.

Hook All sizes.
Thread Red.
Tag Round silver tinsel and yellow floss.
Tail Golden pheasant topping and tippets.
Butt Scarlet wool (black or red ostrich is often used).
Body Light blue floss.
Rib Oval silver tinsel.
Hackle Light blue hackle palmered; blue jay or blue guinea fowl at throat.
Wing Tippet in strands, golden pheasant tail; married strands of scarlet, blue, yellow swan, florican bustard, peacock wing, light mottled turkey tail; married strands of summer duck, and brown mallard; golden pheasant topping over all.
Head Red (some dressers prefer black).

Thunder and Lightning

Another classic pattern, this time having a much simpler wing construction. All the early fly-dressers, including Kelson and Pryce-Tannant, gave this pattern. It was originated, like many other patterns, by James Wright in the mid-nineteenth century, and is still one of the most popular salmon flies in use today. It lends itself well to both hair-wing and tube-fly design.

Hook All sizes.
Thread Black.
Tag Round gold tinsel and yellow floss.
Tail Golden pheasant topping.
Butt Black ostrich.
Body Black floss silk.
Rib Oval gold tinsel.
Hackle Palmered orange cock; blue jay or dyed guinea fowl at throat.
Wing Brown mallard with topping over.
Cheeks Jungle cock.
Head Black.

Em Terror

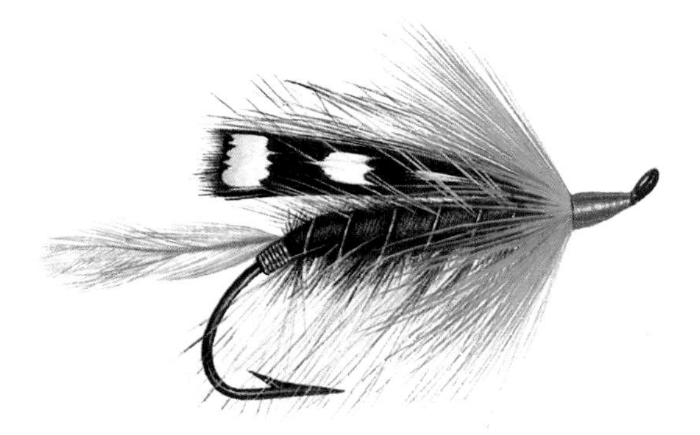

A fairly modern fly from Sweden named after the River Em. It is often used in large sizes, like many Scandinavian patterns; the deep, fast-flowing rivers require such big flies. The Em Terror is used in smaller sizes for sea and lake trout.

Hook All sizes up to 5/0.
Thread Red.
Tail Two orange hackle tips.
Tag Round gold tinsel.
Body Black wool.
Rib Oval gold tinsel.
Hackle Palmered badger cock; blue collar at head.
Wing Two Jungle cock feathers back to back.
Head Red.

Usk Grub

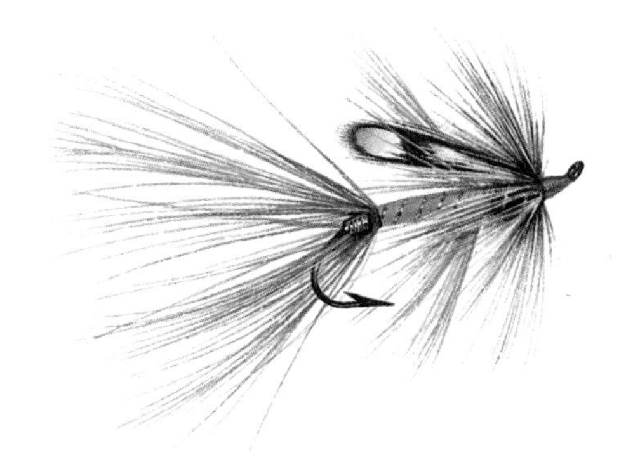

One of the all-time favourite Welsh flies. As its name suggests, it was born, so to speak, on the famous salmon river, the Usk. There were a number of early patterns named the Usk Grub, all slightly different. The dressing given here is now the accepted one. Another pattern, the Welsh Shrimp, is very similar, as are a number of shrimp flies from Ireland, including Curry's Red and Curry's Gold Shrimps.

Hook All sizes up to 1/0.
Thread Red.
Tag Round silver tinsel.
Tail Red golden pheasant body feather wound as hackle.
Body Two halves: orange wool and black wool.
Rib Oval silver or gold tinsel.
Hackle Between the two body colours, a white hackle followed by an orange hackle. At the head, a collar hackle of coch-y-bonddu or badger.
Wing Two Jungle cock hackles back to back.
Head Red.

Kola Fly

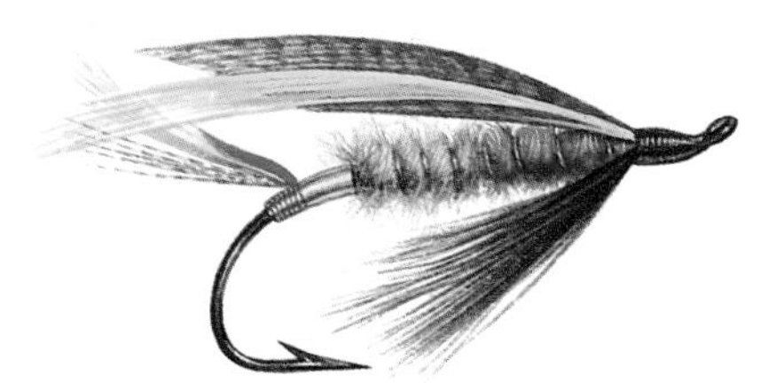

A simple fly from Finland, where there are many excellent fly-dressers. This fly, scaled down, can be used for trout and also seatrout.

Hook All sizes (usually fished in smaller sizes).
Thread Black.
Tag Round silver tinsel and yellow floss.
Tail Grey mallard fibres, red swan, and golden pheasant topping.
Body Two halves: yellow seal's fur, blue seal's fur.
Rib Oval gold over yellow seal's fur; oval silver over blue.
Hackle Black.
Wing Brown mottled turkey tail with golden pheasant over.
Head Black.

Sweep

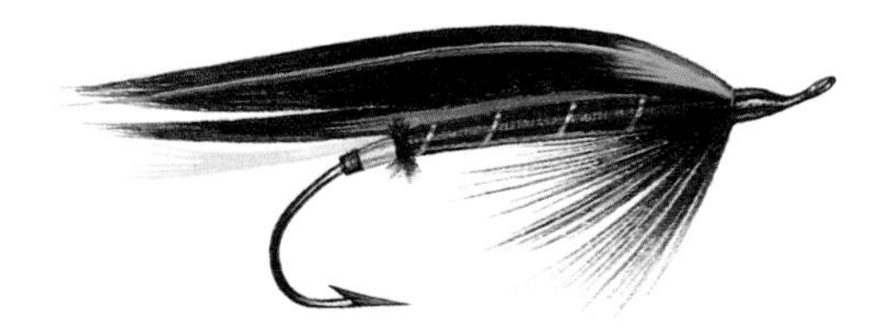

One of many dark or black flies used today, others being the Stoat's Tail, Black Dose, and the Black Bear series. Sometimes a Jungle cock cheek is added, and often the fly is dressed with a black heron hackle.

Hook All sizes.
Thread Black.
Tag Oval silver tinsel and yellow floss.
Tail Golden pheasant topping.
Butt Black ostrich herl.
Body Black silk.
Rib Oval silver tinsel.
Hackle Black.
Wing Black turkey or dyed swan.
Cheek Blue kingfisher.
Horns Blue-yellow macaw (optional).

Skiri Fly

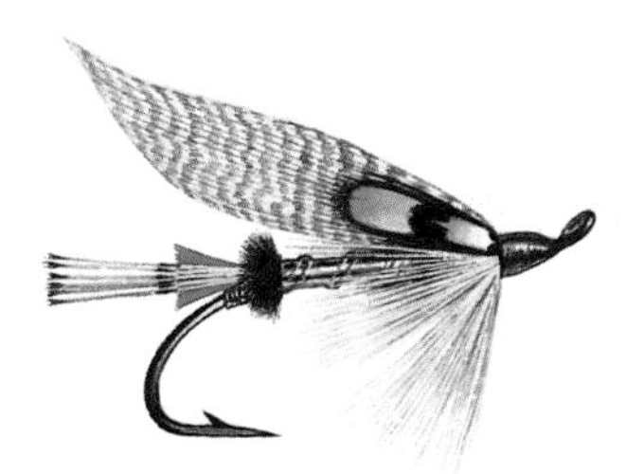

Iceland is justly famed for its salmon fishing. Anglers from all over Europe and America each year make Iceland their fishing Mecca. Many of the flies used in that cold, northern country are variations on established British flies, but the Skiri is an exception.

Hook All sizes.
Thread Black.
Tag Oval silver tinsel.
Tail Golden pheasant tippets and Indian crow (substitute).
Body Flat silver tinsel.
Rib Oval silver tinsel.
Hackle Light grey.
Wing Grey mallard flank.
Cheek Jungle cock.

General Practitioner

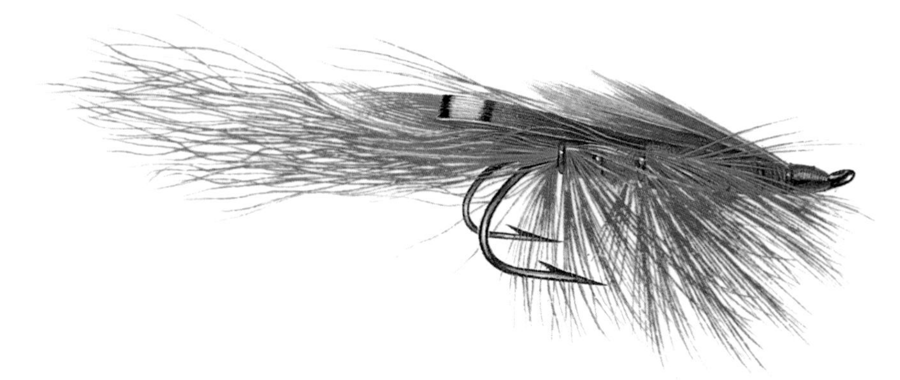

This 'prawn' fly is the creation of Col. Esmond Drury, and though more modern flies constructed out of latex and other such materials have come onto the angling scene, this fly still holds its own and continues to be popular. It received its name from the feathers used in its make-up, namely golden pheasant, abbreviated to GP; hence General Practitioner.

Hook All sizes; best on doubles up to 3/0.
Thread Red or orange.
Tail Orange bucktail (this imitates the prawn's feelers).
Body Orange seal's fur.
Rib Oval gold tinsel.
Hackle Palmered orange cock.
Back Made up of two pairs, each of golden pheasant red body feathers.
Eyes Golden pheasant tippet.
Head Red.

Garry

The Garry, or Yellow Dog as it is sometimes called, was originated by John Wright, son of the renowned James Wright of Sprouston, UK. The original fly used the hair from a golden retriever, hence the name Yellow Dog. This fly is one of the most popular hair-wings and is tied on conventional, double and treble hooks, as well as on tubes and Waddingtons.

Hook All sizes.
Thread Black.
Tag Oval silver tinsel and yellow floss.
Tail Golden pheasant crest (sometimes a few strands of red ibis are added).
Butt Black ostrich (optional).
Body Black floss.
Rib Oval silver tinsel.
Hackle Dyed-blue guinea fowl.
Wing Yellow bucktail over red bucktail.
Head Black.

Butterfly

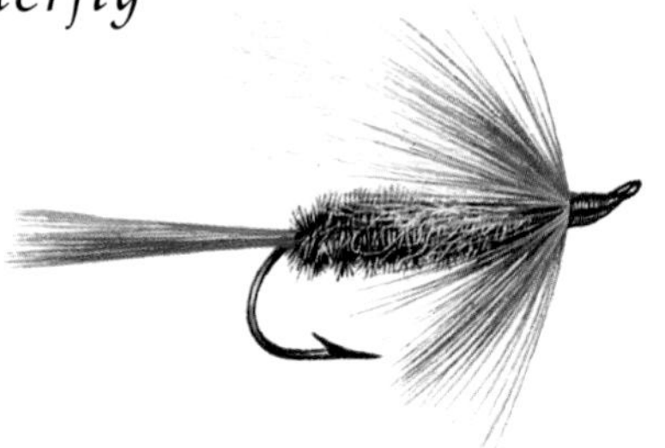

This American pattern is sometimes called Engell's Butterfly, though the originator appears to be one Maurice Ingalls of Fort Lauderdale, Florida. It was devised for use in New Brunswick, Canada. It is sometimes called Ingall's Splay-wing Coachman. The similarity to the Coachman is obvious.

Hook All sizes to 1/0.
Thread Black.
Tail Scarlet hackle fibres.
Body Peacock herl.
Rib None.
Hackle Natural red cock, tied collar style.
Wing Two bunches white calf tail, set at an angle of 45°.

Cosseboom Special

There are a number of flies in the Cosseboom series, but the one given here is the original. All are excellent flies for all species of game fish including, of course, the Atlantic salmon. They are used in all sizes, right up to 5/0 and 6/0 on some waters. Others in the series are the Gold, Red, Yellow and Orange Cosseboom; the Silver Cosseboom is very popular in parts of Newfoundland. Another example, the Indian Cosseboom, is given later.

Hook All sizes.
Thread Red.
Tag Oval silver tinsel (sometimes embossed tinsel is used).
Tail Tuft of olive floss.
Body Olive green floss.
Rib Oval silver tinsel (embossed tinsel sometimes given).
Hackle Bright yellow collar hackle.
Wing Grey squirrel.

Munro's Killer

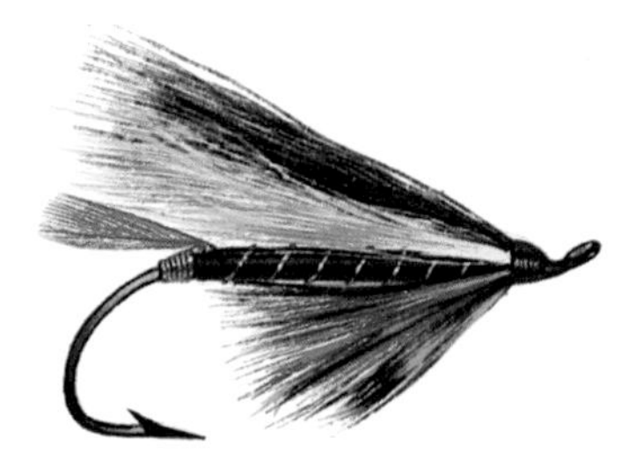

One of the most popular British hair-wings of recent years. The example given has a wing of yellow hair with black over. Other dressings give a black over brown wing, and yet a third uses a yellow-dyed squirrel. All appear to catch their fair share of fish.

Hook All sizes.
Thread Black.
Tag Oval gold tinsel.
Tail Orange hackle tip.
Body Black floss.
Rib Oval gold tinsel.
Hackle Orange cock, blue-dyed guinea fowl in front (or blue jay).
Wing Black squirrel over yellow squirrel.
Head Black.

Black Bear Green Butt

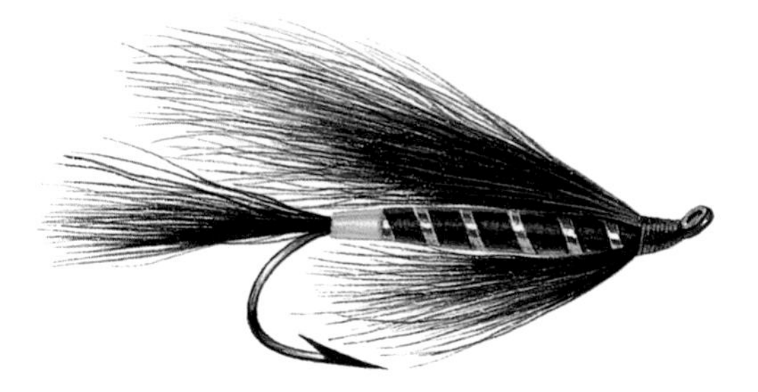

This fly is just one of a series of American hair-wing flies for Atlantic salmon. All are the same except for the butt, which is green in this pattern. Others are the Red Butt, Yellow Butt and Orange Butt. All are a development of the original Black Bear, a fly of the 1920s which was plain and unadorned.

Hook All sizes.
Thread Black.
Tail Black bear hair.
Butt Fluorescent green floss.
Body Black floss.
Rib Flat silver tinsel.
Hackle Black cock.
Wing Black bear.

Indian Cosseboom

A recent Canadian version of the Silver Cosseboom, it is used on rivers in the New Brunswick area. The long red butt on the fly is virtually the only difference between it and the original Silver Cosseboom.

Hook All sizes (sometimes tied on long shank streamer hooks).
Thread Black.
Tag Oval silver tinsel.
Tail Golden pheasant crest.
Body Two parts: first quarter, red floss; rest, flat silver tinsel.
Rib Oval silver tinsel.
Hackle Yellow collar hackle.
Wing Grey squirrel.

Grey Monkey

A hair-wing version of an old feather-wing pattern. The fur originally used on the body was in fact monkey fur. Grey seal or any grey fur makes a good substitute.

Hook All sizes.
Thread Black.
Tail Teal fibres.
Body Two halves: yellow floss followed by grey seal's fur.
Rib Oval silver tinsel.
Hackle Badger.
Wing White bucktail.

Naranxeira

B. Martinez of Pravia, Asturias, Spain, is noted throughout Europe for the high quality of his salmon flies. Most Spaniards prefer to fish with the more ornate feather creations, and this fly is one of the few hair-wing flies that Senor Martinez ties.

Spain is the southern limit of the Atlantic salmon.

Hook All sizes to 5/0.
Thread Black.
Tag Oval silver tinsel and yellow floss.
Tail Golden pheasant crest and barred teal.
Butt Black wool.
Body In two halves: orange floss and black floss.
Rib Oval silver tinsel.
Hackle Two: a pink hackle tied before the wing, and a grey hackle tied after.
Wing Grey squirrel.

Rusty Rat

The Rat series of American salmon flies are probably the best-known flies from the USA. Two examples are given in this book, the Rusty and the Grey. Others include Copper, Black, Red, Silver and Gold. All use grey fox for the wing; when this is not available, grey squirrel is often used. These patterns are used throughout the world.

Hook All sizes.
Thread Red.
Tag Oval gold tinsel.
Tail Green peacock sword fibres.
Body In two halves: yellow floss followed by peacock herl. The yellow body is veiled under the wing with a length of floss (sometimes orange floss is given).
Rib Oval gold tinsel (optional).
Hackle Grizzle collar hackle.
Wing Grey fox or squirrel.
Head Red.

French Salmon Fly

This fly comes from the firm of Guy Plas and, like his excellent trout patterns, it uses the sharp, glass-like feather fibres from the spade hackles of cockerels found on both sides of the Pyrenees. The feathers from the famous coqs de Leon are renowned throughout the fly-fishing world. This is one of a series of similar flies, some having different body colours and subtle differences in feather shading.

Hook 2-8.
Thread Black.
Tag Oval silver tinsel and red floss.
Tail Grey mottled cock hackle fibres.
Body Black silk.
Rib Oval silver tinsel.
Hackle Grey mottled cock hackle fibres.
Wing Black squirrel with grey mottled fibres over.

Heggli

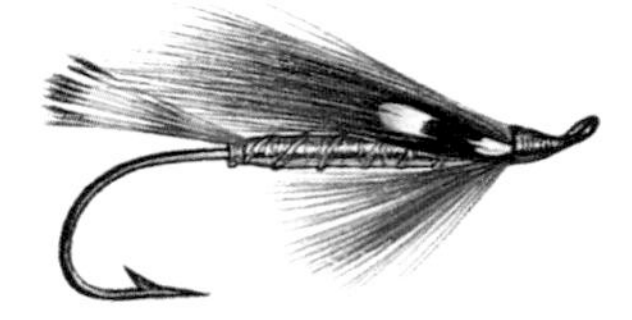

This Norwegian 'all-round' fly is shown as a low-water, hair-wing salmon fly. The fly is, in fact, used for all species of game fish-trout, seatrout and salmon. The original fly uses brown mallard for the wing; the Jungle cock cheek is optional.

Hook All sizes (low-water style illustrated).
Thread Black.
Tag Oval silver tinsel.
Body Flat silver tinsel.
Rib Oval silver tinsel.
Hackle Natural dark red cock.
Wing Brown squirrel (feathered fly uses brown mallard).
Cheek Jungle cock.
Head Black.

Tosh

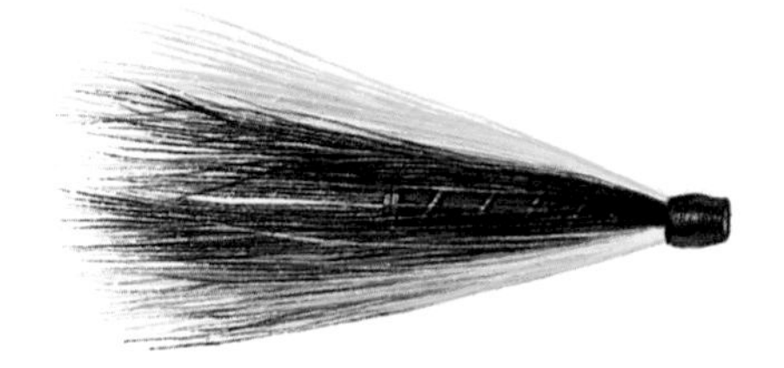

This is a modern British tube fly, used on the Spey. It is sometimes called the Spey Tosh or plain Yellow and Black. It is used in all sizes to over 3 in (7.5 cm). Like all good Scottish hair-wing flies, the original was supposed to have been tied with the hair from a black dog and a golden dog, which leads me to suspect that there must be many bald dogs in Scotland, for it is an extremely popular fly.

Tube 1-3 in (2.5 -7.5 cm).
Thread Black.
Tail None.
Body Black floss silk.
Rib Oval gold tinsel.
Hackle None.
Wing Black/yellow/black/yellow bucktail.
Head Black.

Akroyd

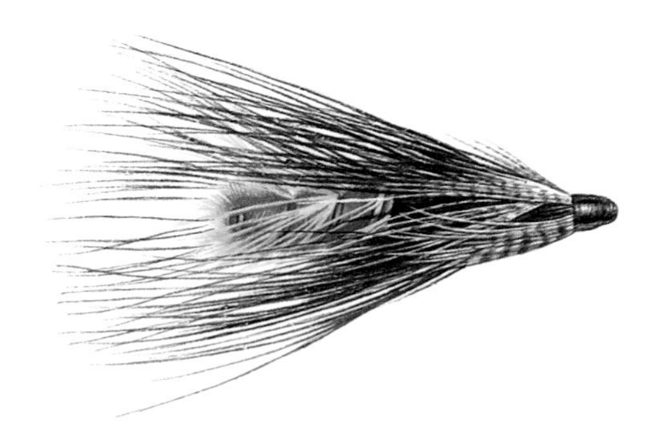

This is a tube version of the famous Akroyd, a Dee strip salmon fly created in the late nineteenth century by Charles Akroyd of Brora. The noticeable features of the Dee-type flies were the long, narrow single-feather wings and the use of long heron hackles. I have used a heron hackle in the tube dressing; long black bucktail could be used as an alternative.

Tube 1-3 in (2.5-7.5 cm).
Thread Black.
Tail None.
Body Orange seal's fur followed by black seal's fur.
Rib Oval gold tinsel and palmered yellow cock hackle.
Hackle Teal.
Wing Long black heron hackle fibres, or black bucktail.
Head Black.

Jock Scott

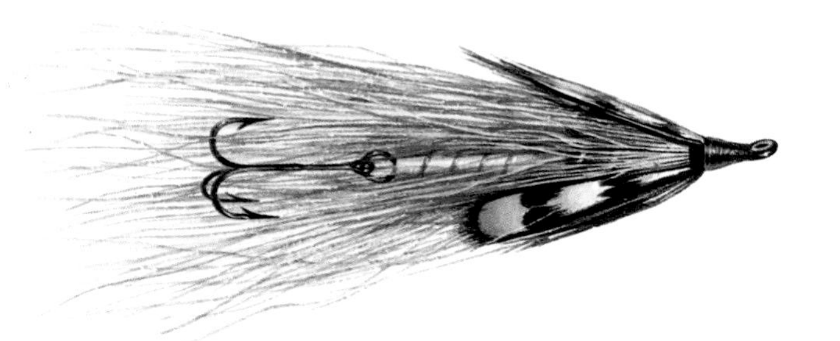

The Jock Scott is amongst the best known salmon flies still in use today. Where Atlantic salmon run, somebody will be using a Jock Scott no matter what country. It is a truly classic fly. The pattern here is tied on a Waddington Elverine shank, and is an alternative form to the normal tube fly. Like tube flies, and to a certain extent hair-wing versions of standard feather patterns, the dressing ingredients are the choice of each individual tyer. There are no hard-and-fast rules for the wing composition.

Hook ½-3 in (1-7.5 cm) Waddington shank.
Thread Black.
Tail None.
Body Yellow floss followed by black floss.
Rib Fine silver oval over yellow floss; wider over black.
Hackle Guinea fowl.
Wing Mixed fibres: red, blue, yellow and brown bucktail.
Cheeks Jungle cock.

Hairy Mary

Most people would be flattered to have a fly named after them, but I doubt it pleased Mary, who ever she was. The fly came to light in the early 1960s, created by John Reidpath of Inverness. There can be few anglers who have not heard of this hirsute female fly.

Hook All sizes.
Tag Oval gold tinsel.
Thread Black.
Tail Golden pheasant crest.
Body Black floss or wool.
Rib Oval gold tinsel.
Hackle Bright blue hackle.
Wing Brown bucktail.

Blue Mary

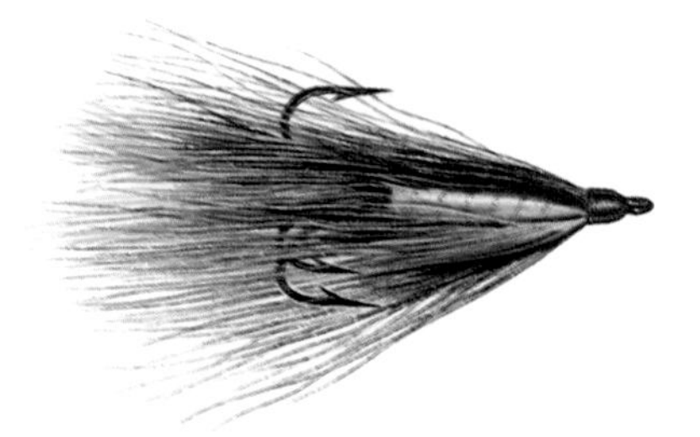

This simple variation of the well-known Hairy Mary accounted for an 8 lb (3.6 kg) salmon on the first cast on the first occasion it was used. The example shown is tied on a long-shank, Drury-type black treble. These hooks do not make pretty flies, but they do make good hooking flies.

Hook Treble long shank salmon; all sizes.
Thread Black.
Tail Yellow golden pheasant body feather fibres.
Body Blue floss.
Rib Silver oval tinsel.
Hackle None.
Wing Alternate tufts of black and brown bucktail.

Grey Rat

The original and most effective of the Rat series of American hair-wing salmon flies. The flies go back to the early part of the twentieth century, and originated in New Brunswick, Canada, where they are still used today. A full history of the Rat series is given by Col. Joseph D. Bates in his authoritive book, *Atlantic Salmon Flies and Fishing*.

Hook All sizes.
Thread Red.
Tag Flat gold tinsel.
Tail Golden pheasant crest feather.
Body Grey fox under fur.
Rib Flat gold tinsel.
Hackle Grizzle collar hackle.
Wing Grey fox guard hairs (substitute squirrel if not available).
Head Red.

Red Abbey

A traditional North American pattern going back to before World War I. It was used extensively in Quebec. It was an adaptation of an early fancy trout pattern.

Hook All sizes.
Thread Red or black.
Tag Oval silver tinsel (sometimes embossed tinsel is used).
Tail Strip of red-dyed goose or swan.
Body Red floss silk.
Rib Flat silver tinsel (sometimes embossed).
Hackle Collar hackle of natural red cock.
Wing Brown squirrel or brown bucktail.
Head Red or black.

Squirrel and Orange

This fly is just one of a range of Squirrel flies, the orange being the most popular. Others include the Squirrel and Black and Squirrel and Gold.

Hook All sizes.
Thread Black or red.
Tail Golden pheasant crest.
Body Flat golden tinsel.
Rib Oval gold tinsel.
Hackle Orange cock.
Wing Grey squirrel.
Head Red.

Dry Flies

Pink Lady

A pattern devised by the American, pioneer dry-fly salmon fisherman, George M. L. La Branche. The Pink Lady, along with La Branche's other patterns, are considered to be some of the earliest salmon dry flies, created in the early 1920s.

Hook Up-eyed light wire salmon dry-fly hook 8-4.
Thread Black.
Tail Light red (ginger) cock hackle.
Body Pink floss.
Rib Oval gold tinsel.
Hackle Light red (ginger) palmered with yellow collar at the front.
Wing None.

Colonel Monell

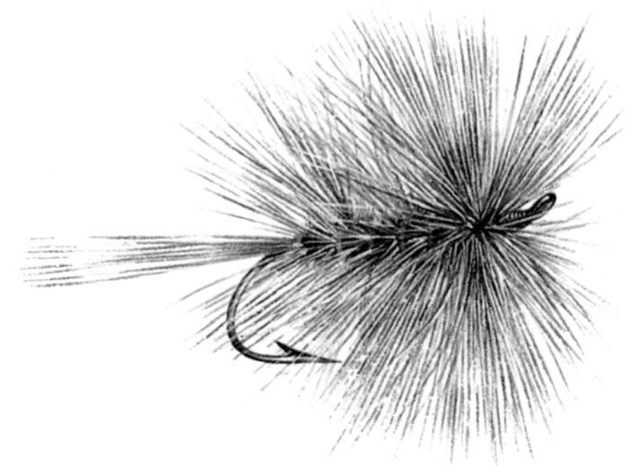

Another of La Branche's patterns still used today. It was named after his fishing companion, Col. Ambrose Monell, who appears to have taught La Branche the unique art of dry-fly fishing for salmon.

Hook Up-eyed light wire salmon dry-fly hook 8-4.
Thread Black.
Tail Grizzle hackle fibres.
Body Peacock herl.
Rib Red silk.
Hackle Palmered grizzle hackle.
Wing None.

Whiskers

A modern dry fly for salmon, popular in Newfoundland and New Brunswick. It is very similar to the Wulff series of dry flies, which are also used for salmon.

Hook Up-eyed light wire salmon dry-fly 6-4.
Thread Black.
Tail Grey squirrel.
Body Black wool.
Rib None.
Hackle Palmered brown cock hackle.
Wing A bunch of deer body hair, fan-shaped on top of the hook, sloping slightly forward.

Buck Bug

A fly that uses the buoyant property of deer hair, which must be clipped roughly for best effect. There are many deer hair flies of similar appearance used for the salmon; the Rat-faced McDougal and the Irresistables are further examples.

Hook Up-eyed fine wire salmon dry-fly hook 8-4.
Thread Black.
Tail Fox squirrel fibres.
Body Clipped deer hair roughly trimmed.
Rib None.
Hackle Palmered brown cock with a further hackle tied at the front.
Wing None.

Bass flies

The two main black bass species are the large-mouthed bass (*Micropterus salmoides*) and the small-mouthed bass (*Micropterus dolomieu*). Both are widely distributed throughout the United States of America and are considered to be the most sporting of fish. They have been introduced successfully into some of the dams of South Africa, and are providing good sport in Spain. They can easily be distinguished one from another, as one would suspect from their names, by their mouths. The large-mouthed bass has an upper jaw extending beyond the eye; the mouth of the small-mouthed bass ends just below the eye. Although the two species can be found in the same water, as a general rule the large-mouthed is a creature of the warmer waters of muddy lakes and ponds, and slow sluggish rivers. The small-mouthed, on the other hand, prefers large clear lakes and faster-flowing rivers. Many bass fishermen can tell which species they have on the end of their lines from the tenacity of the fight. The small-mouthed is thought to be the scrappier fighter of the two.

Both species are fearsome predators, feeding on a wide variety of creatures: fish, frogs, birds and, when young and smaller, a wide range of insect life. Even large bass can be tempted to come up to the surface to take creatures such as bees and beetles.

The best-known artificial lure for these fish is, without doubt, the Popping Bug in its various forms, made either from balsa, cork, plastic, or from clipped deer hair. The purpose of the popper is to make the quarry take notice by causing a disturbance on the surface. Some poppers are made to dive deep on retrieve; some pop and blurp across the surface in little shallow hops; others are shaped to cause a wake. Some poppers are devised with natural creatures in mind, such as frogs or bumble bees; others, the bright and garish attractors, look as though they have come straight out of Disneyland.

Bass are also taken on conventional wet and dry flies, and in particular on large nymphs tied to represent dragonfly larvae, helgrammites and other big aquatic nymphs. Flies tied to represent crayfish are also a good bet for the black bass.

Streamer and bucktail flies figure in the armoury of the bass fly-fishermen. Because of the nature of the habitat – bass favour weedy water – many of these lures are tied either as weedless flies (see the Black Leech pattern), or they are tied on large Keel hooks.

In recent years flies and poppers have been made from the versatile deer hair. Many fishermen prefer this type of fly because it is easier to cast. However, some of the early deer hair flies were pretty basic; for instance the Powderpuff, a ball of undyed and unclipped deer hair looking for all the world like a tiny, rotund hedgehog.

A number of American fly-tying books feature several bass patterns. There is one book that deals with flies and lures for the bass and other panfish, *Fly Tying and Fly Fishing For Bass and Panfish* by Tom Nixon (A. S. Barnes & Co.) published in 1968. It is well worth reading.

Hare Muddler

Yet another version of the ubiquitous Muddler Minnow. This version dispenses with the usual clipped deer hair, and uses hair from the hare's mask for both the head and the collar. The fly is less buoyant than the original muddler, allowing it to get down to the fish much quicker and yet retaining the Muddler magic.

Hook Long shank 10-4.
Thread Black or brown.
Tail Oak turkey.
Body Flat gold tinsel.
Rib Gold oval tinsel (optional).
Hackle A collar of dubbed fur from a hare's face.
Wing Grey squirrel flanked by oak turkey slips.
Head Dubbed hare's mask fur.

Zonker

One of the killing lures of recent years for both trout and bass. The fly owes a lot to the New Zealand Rabbit lures. The Zonker is not normally tied Matuka-style, but the wing is tied down at the head and tail. A little strong glue applied across the back before securing the rabbit fur prolongs the life of the fly.

Hook Long shank 8-4.
Thread Red.
Tail A slip of white rabbit fur; this is taken over the back of the fly.
Body Gold or silver mylar tubing over a wool or silk underbody.
Rib None.
Hackle Scarlet cock hackle fibres.
Wing Provided by the rabbit fur taken over the back.

Black Prisma Shiner

Dave Whitlock's Shiner patterns have been used for both trout and bass with great success. I have based this black version on such flies.

Hook Long shank 8-4.
Thread Black.
Tail None.
Body Silver prismatic sheet.
Rib None.
Hackle None.
Wing Black marabou flanked by two grizzle hackles with peacock herl over.
Eyes Plastic goggle doll's eyes.

Red Bi-visible

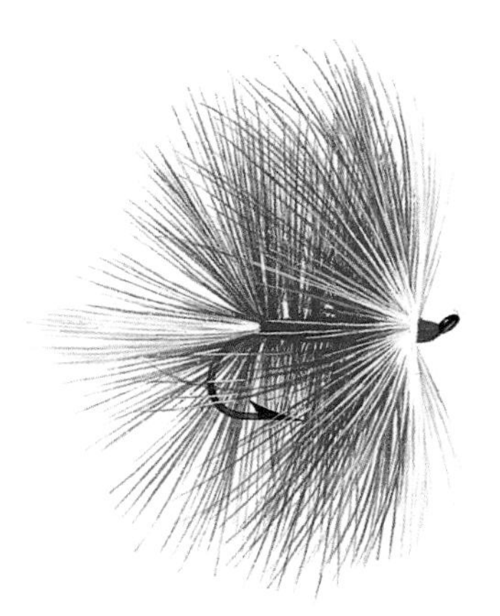

A traditional, conventional fly. The colour red makes it particularly attractive. Other colours are yellow, with a black front hackle and tail; and olive, with the usual white front hackle and tail.

Hook 8-6.
Thread Red.
Tail White hackle fibres.
Body Palmered red cock hackle.
Rib None.
Hackle White cock hackle.
Wing None.

Black Leech

This fly is very similar to, and sometimes confused with, the popular lure called Black Bugger. The main difference is that this pattern is eyed. The Leech can also be tied in white, brown and olive and is recommended for trout.

Hook 1/0 salmon hook.
Thread Black.
Tail Black marabou.
Body Black wool.
Rib Palmered black cock hackle.
Hackle None.
Wing None.
Eye Glass; white, red or amber.

Keel Bass Bug

One of a whole series of different coloured, weedless Keel-hook flies. The one depicted here is a red version. Others include natural, yellow, black, olive and bright green.

Hook Keel hook 4-6.
Thread Red.
Tail Two grizzle cock hackles with red deer hair.
Body Dyed-red deer hair.
Rib None.
Hackle None.
Wing Unclipped red deer hair.
Head Clipped deer hair.

Poppers

One of the most popular lures for the large- and small-mouthed bass. Poppers come in many forms; some dive while making distinct plopping noises; others cause maximum activity on the surface; and some wake across the surface when retrieved. All prove very attractive to many species of fish. I often use them for still-water trout.

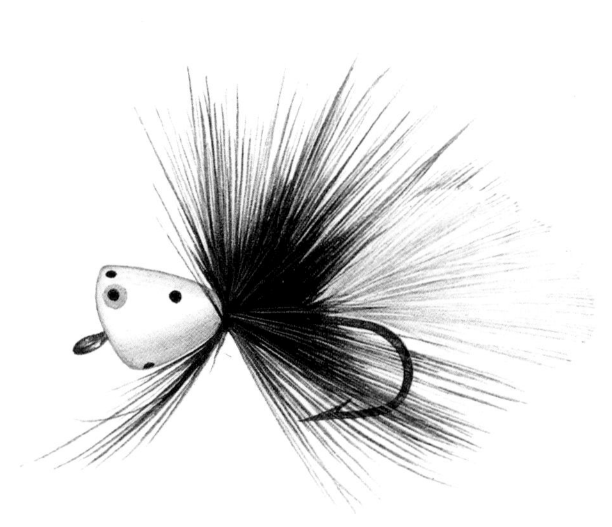

Black and White
Hook All sizes to 1/0.
Thread Black.
Tail Black and white hackles.
Body None.
Rib None.
Hackle Black cock hackle.
Wing None.
Head White balsa or cork with painted eye.

Yellow and White
Hook All sizes to 1/0.
Thread Black.
Tail Yellow cock hackles.
Body None.
Rib None.
Hackle Yellow.
Wing None.
Head White, with painted eye.

Deer Hair Bass Bug

A clipped deer-hair version of the popping bug. Many fishermen prefer the softer, lighter deer bugs to more conventional balsa or cork versions.

Hook Long shank 6-2.
Thread Black.
Tail Two grizzle hackles.
Body Clipped black deer hairs.
Rib None.
Hackle None.
Wing None.

Frog

There is nothing that appeals to a bass more than a juicy frog as it swims across the water's surface. Imitations of frogs can be created, like this one, out of deer hair clipped to a frog shape, or out of balsa or cork in the conventional popping bug style.

Hook 8-4.
Thread Olive.
Tail None.
Body Olive/green-dyed deer body hair clipped to shape.
Rib None.
Hackle None.
Legs Dyed bucktail.

Bullet Frog

A simple deer-hair bug. Many bass fishermen prefer unadorned bugs of this type, thinking them easier to cast than those with eye and rubber leg embellishments. The bass do not seem to care either way.

Hook Long shank 8-4.
Thread Olive.
Tail Two bunches of olive bucktail.
Body Olive bucktail tied at head, then taken down the shank and tied at tail. The ends form the tail.
Rib None.
Hackle None.
Wing None.

Saltwater flies

The sport of fly-fishing in saltwater is still in its infancy. Though it is practised in many parts of the world, it has not yet attracted the number of fishermen that other forms of fly-fishing have.

The use of a feathered hook is no new thing. For centuries, fishermen have sought traditional quarry by means of one lure or another. The very basic mackerel feathers used on hand lines, the larger feathered hooks used when jigging for cod and similar species, the crude, feathered lures used by professional tuna fishermen, are all artificial 'flies', albeit rather primitive ones.

It is to the warmer waters of the world that we must look for the best in saltwater fly-fishing sport: the waters around the coast of the Southern States of the USA, Central and South America, the West Indies, South Africa, the islands of the Pacific, and Australasian waters. All these areas hold a wealth of fish that can be taken on an artificial fly, whether by normal fly casting, jigging or trolling.

To describe some of the creations used to take such fish as 'flies' is something of a misnomer. At the smaller end of the range they are devised to imitate shrimps and other crustacea; larger lures are tied to represent various baitfish that make up the quarry's diet; and a third type are attractors.

For most saltwater fly-fishing, the so-called flies need not be over elaborate as long as they have some in-built life provided by the materials used in their construction. Long hackles that breathe and pulsate in the water and non-tarnishing tinsels that add flash and glitter are essential for good killing flies. To the fish, lures must look like food; they must look alive and they should move in such a way as to make the fish take notice, turn and seize. Flies such as the Joe Brook's Blonde series, any of Lefty Kreh's, or some of the modern Dave Whitlock creations, have these characteristics. All the angler provides is a little skill and a lot of patience.

The world is getting smaller, and anglers from colder climes are now seeking their sport in warmer waters. They may seek bonefish in the blue waters of the Bahamas, or tarpon in Mexico or Florida. And once bitten they are bound to return.

In the colder waters around the British coast, nineteenth-century fly-fishermen sought bass with flies like the Alexandra trout fly and other trout and salmon patterns. They caught pollack on red and white flies called Cuddy flies named after the small coalfish which the pollack fed upon. Another traditional British fly was the Shaldon Shiner, a blue-coloured fly which was considered excellent for bass. Even the soft-mouthed grey mullet had a fly tied to tempt it; a lure which used owl feathers and white wool, and was often tipped with a sliver of shellfish.

Those who wish to take their fly rods to saltwater would do well to read *Fly Fishing in Salt Water* by Lefty Kreh, published in the mid 1970s. It is considered to be the definitive work on the subject.

Rainbow Mohawk

A multi-coloured fly for many species of sea and estuary fish. This pattern and others in the series was devised by Barry Kent during his time in South Africa. The fly was created for the North American market with various Pacific salmon in mind.

Hook Long shank, all sizes.
Thread Black.
Tail Red bucktail.
Body As tying thread.
Rib None.
Hackle None.
Wing Bucktail: individual tufts, mauve, blue, green, yellow, orange, red.

Red Mohawk

Another in the Mohawk series, in smaller sizes. I have found these flies to be an excellent lure for stillwater trout. Green, blue and yellow versions are tied in the same way with the gradual variance of colour in the wing.

Hook Long shank, all sizes.
Thread Red.
Tail White bucktail.
Body As tying thread.
Rib None.
Hackle None.
Wing Bucktail: individual tufts, black or red, light red, pink.

Cockroach

This fly was devised by the well-known exponent of salt-water fly-fishing. Lefty Kreh. Lefty describes this fly as a basic Tarpon pattern. It should not exceed 4 in (10cm) in length.

Hook 3/0-5/0 ring eyed.
Thread Black.
Tail None.
Body None.
Rib None.
Hackle Natural brown deer hair.
Wing Four grizzle hackles, shiny side outside.

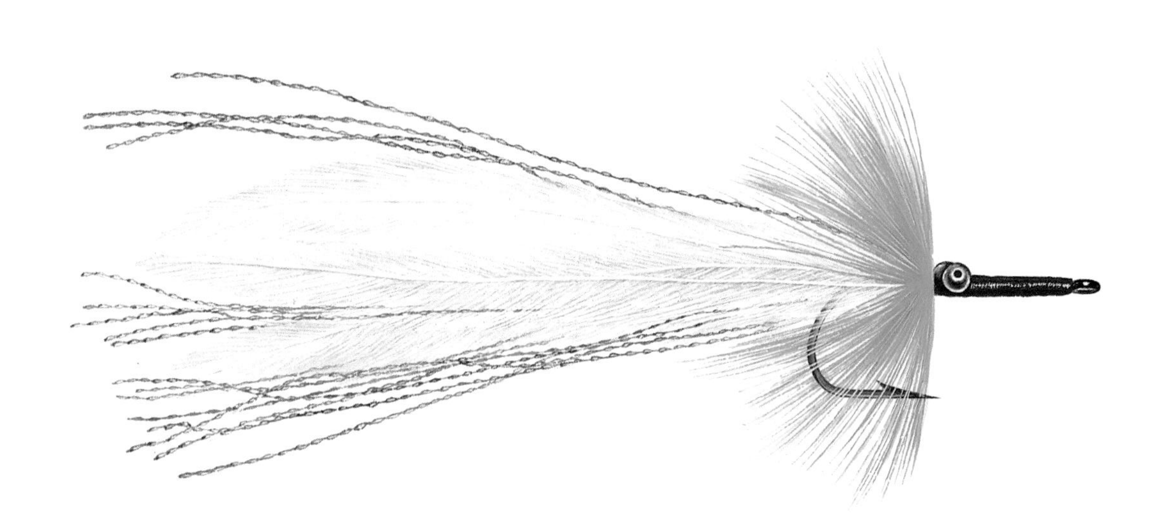

Tarpon Streamer

I based this pattern on the usual standard Tarpon type fly, with the addition of a pair of weighty bead eyes in order to impart a diving action to the lure. It can be tied in a variety of colours.

Hook 3/0-5/0 ring eyed.
Thread Red.
Tail White saddle hackles with silver tinsel (mylar, flashabou or shimma).
Body Red; a long extension of the head.
Rib None.
Hackle Blue.
Wing None.
Eyes Bead chain (optional).

Platinum Blonde

This is one of a series created by the late Joe Brooks, who took his fly rod to all corners of the fishing world and sought his quarry in both fresh and saltwater.

Other flies in the 'Blonde' series are:
Strawberry Blonde Red wing/orange tail.
Honey Blonde Yellow wing/yellow tail.
Pink Blonde Pink wing/pink tail.
Black Blonde Black wing/black tail.
Argentine Blonde Blue wing/white tail.

The silver body is constant in all patterns. The flies are tied in a wide variety of sizes for many species of saltwater fish. They have also been used for both trout and salmon.

Hook 2-3/0 ring eyed.
Thread White.
Tail White bucktail or other white hair.
Body Flat silver tinsel.
Rib Oval silver tinsel (optional).
Hackle None.
Wing White bucktail or other hair.

Permit Fly

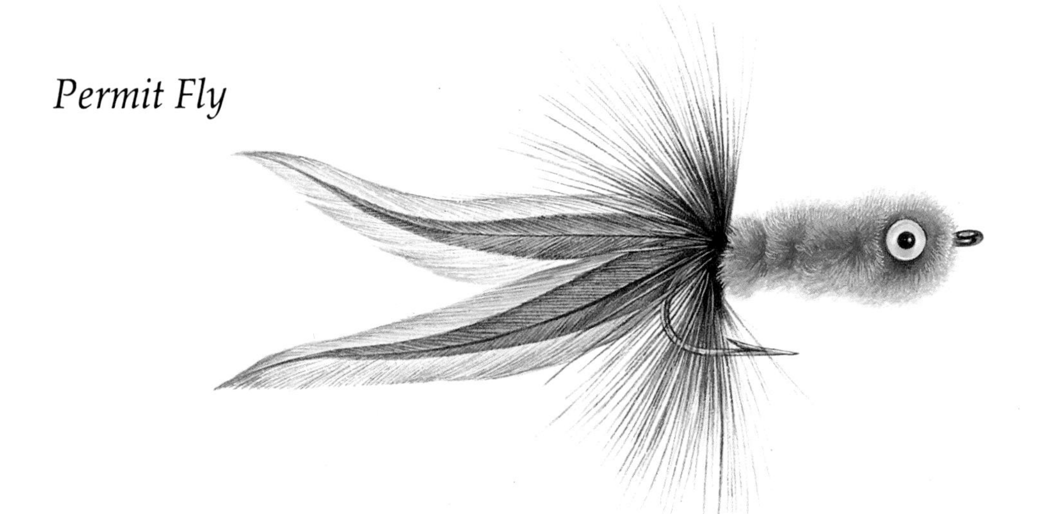

A lure devised to take the sporting fish of the Florida Keys. They are also taken on the usual bonefish fly patterns.

Hook 4-1/0 ring eyed.
Thread Tan.
Tail Furnace hackle either side of badger hair.
Body Tan chenille.
Rib None.
Hackle Furnace hackle.
Wing None.
Eyes Amber glass eyes.

Lefty's Deceiver

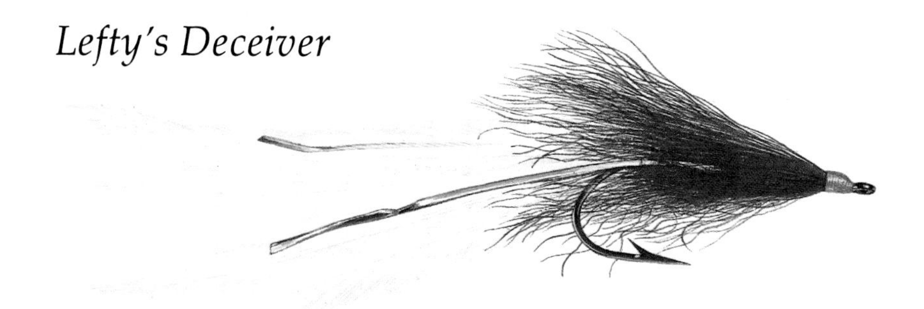

Another of Lefty Kreh's standard patterns, this is used for a wide variety of saltwater game fish, and can be tied in a number of different colour combinations.

Hook 3/0-5/0.
Thread White.
Tail White saddle hackles and strands of silver mylar.
Body None.
Rib None.
Hackle Dyed red bucktail.
Wing None.

Sea-ducer

A fairly recent pattern created by Chico Fernandez of Miami, Florida. Like most of the sea lures, it can come in a wide variety of colour permutations. It is used for such fish as sailfish, amberjack, redfish, tarpon and dolphin.

Hook 3/0-5/0 ring eyed.
Thread Black or red.
Tail 4-6 yellow and grizzle, saddle hackles, plus silver strands (mylar, flashabou, and so on).
Body Palmered yellow hackle.
Rib None.
Hackle Red hackle.
Wing None.

Skipping Bug

The example depicted here is the black/gold. Others in the series are an all-yellow, blue/ silver, green/silver, a blue and white, and red and white.

Hook 1/0-2/0.
Thread Black.
Tail Yellow bucktail, black bucktail over.
Body Balsa/cork painted black on top, gold beneath.
Rib None.
Hackle None.
Wing None.
Eye White with red pupil.

Barracuda Streamer

There are many who fear the multi-toothed baracuda more than the shark. This fly was devised especially for this fearsome fish, which puts up an exciting fight on a fly rod.

Hook 3/0-5/0.
Thread Red.
Tail Four white saddle hackles, plus strands of silver mylar (or flashabou).
Body None.
Rib None.
Hackle Olive green bucktail.
Wing None.

Horror

This is a well-known bonefish pattern, the brain-child of Pete Perinchief of Bermuda. It is tied upside-down to prevent snagging up on bottom seaweed.

Hook 2-8 (tied upside down).
Thread Red.
Tail None.
Body Yellow chenille.
Rib None.
Hackle None.
Wing Brown bucktail.

Hagen Sand's Bonefish Bucktail

This fly has become a traditional saltwater pattern. It is named after its creator, Hagen R. Sands of Key West, Florida, USA.

Hook 1-1/0.
Thread Black.
Tail None.
Body None.
Rib None.
Hackle White bucktail.
Wing Yellow and grizzle hackles.
Eye Yellow with black pupil.

Appendices

GLOSSARY OF FLY-TYING MATERIALS AND TERMS

Advanced wings Dry-fly wings tied so that they tilt over the eye of the fly.
Attractor flies Flies tied with no specific creature in mind but having plenty of flashy tinsel, etc., in order to attract the fish.
Biot Narrow side of a goose primary feather.
Bi-Visible North American dry-fly type, designed for high visibility with the use of contrasting-coloured hackles.
Bucktail flies Lures for most species of game fish, tied with the hair of various animals, such as deer, squirrel, goat and calf.
Built wings Wings for salmon flies made from a series of married feather fibres.
Bunch wings Wings for dry flies made from hair, usually divided into two bunches and tied upright or spent, as the pattern dictates.
Butt The part of the salmon fly behind the tail; usually of ostrich herl or wool.
Cape(s) Skins of feathers taken from the necks of domestic fowl and other birds.
Coverts Feathers taken from the base of a bird's wing. Those on the outside of the wing are the marginal and lesser coverts; on the inside, under-coverts.
Crests Feathers found on the top of the head of some birds. In the golden pheasant they are often called toppings. They are used for tails, and also for veiling salmon wings.
Dapping fly A heavily-hackled dry fly that dances on the water surface. Used mostly on stillwater for trout and seatrout.
Dee wing flies Salmon flies used on the River Dee in Scotland having thin strips of brown mallard for the wings.
Detached body The body of a fly that extends beyond the hook. Usually formed separately from the fly and tied onto the hook shank before the rest of the fly is tied.
DFM Daylight fluorescent material.
Dry fly A fly created so that it floats on the surface of the water.
Dubbing Any fur, natural or artificial, spun onto waxed thread then wound onto the hook to form the body, etc, of a fly.
Dun The sub-imago of Ephemeroptera, referred to by anglers as the 'dun'. The imitation of the said sub-imago. A mousey-brown colour.
False hackle A few fibres of hackle tied beneath the hook, and not wound around it as in a conventional hackle. Sometimes called a beard hackle.
Fan wings Wings formed on flies such as mayflies, using two duck breast or other feathers. They sit like two small fans on top of the hook.
Fan-wing flies Flies tied using the 'fan wing' method.
Flue The fluffy fibres or herls that stand out on individual feather fibres. Ostrich and peacock are good examples of feathers that have flues.
Fore-and-aft flies Flies, usually dry flies, that have a hackle at the head and at the tail.
Grub flies Salmon flies of the shrimp variety.
Grub hook A special hook made by Partridge's of Redditch, used for tying pupae flies, shrimps and some Italian patterns.
Hackles The feathers used around the hook to simulate the legs, etc. The individual hackles

usually come from the neck region, but other feathers, such as those from game birds, are used.

Hackle fibre wing Similar to the bunch wings, but hackle fibres are used instead of hair.

Hackle point Tip of the hackle.

Hackle point wings Dry-fly wings made from two or more hackle tips.

Hair-wing flies Salmon flies that use a wing of various hair instead of the normal feathers.

Herl Feather flue, such as ostrich or peacock fibres; other herls are used for the bodies of flies. Usually taken from wing quills, goose, swan and, in the past, condor.

Irons Hooks; usually refers to salmon hooks.

Keel hooks An upside-down hook devised for tying 'weedless' flies, originated by Dick Pobst. Large sizes are used for trout and bass lures; small sizes for natural-looking dry mayfly patterns.

Larva The nymphal stage of some aquatic insects.

Low-water flies Lightly-dressed salmon flies for low-water summer fishing. The dressings of such flies are very much abbreviated.

Lures Large flies of the streamer and bucktail variety used to imitate small baitfish, or tied as attractors.

Married fibres Feather fibres taken from different birds and/or colours joined together to form a multi-coloured wing. Used mainly on built-wing salmon flies and some seatrout patterns.

Matched wing Slips of feather taken from a left and right side. Quills are matched for size, colour and curvature.

Mixed wing Refers to the wing of certain salmon flies. Much the same as built wing.

Old English Bait hook A hook made by O. Mustad & Son, ideal for nymphs and pupae.

Nymphs The larval stage of some aquatic insects. Artificial flies tied to imitate such larvae. Any other aquatic insects such as corixae, shrimps, bugs, etc., are termed nymphs.

Optic flies Flies used for trout, steelhead, bass and some sea fish. Have large, exaggerated eyes.

Palmer fly A fly with a hackle wound from head to tail, referred to as palmered. The Woolly Worm is an example of a palmered fly.

Palmer hackle See palmer fly.

Parachute fly A fly with the hackle wound on the horizontal plane, not vertically as in conventional flies.

Popper Flies that use cork, balsa or plastic for the heads. Sometimes they are constructed out of clipped deer hair. They are used for both large- and small-mouthed bass as well as trout. They make a distinct popping noise when retrieved and their purpose is to cause a commotion on the surface.

Pupa The third stage in the life cycle of some aquatic insects, such as chironomid midges, sedges, and so on.

Quill A feather from the wing or tail.

Quill body A body made from a hackle stalk, or any other feather fibre with the flue scraped off.

Rib A tinsel or any other medium tied around a fly's body in order to simulate the natural segmentation of an insect's body, to add flash and glitter to the fly, to protect delicate feather fibres from the trout's teeth, and in some circumstances to add a degree of weight to the fly.

Ribbing tinsel Either gold, silver or copper; used to provide the rib on a fly.

Saddle hackle Feather found at the top of a cockerel's rump; used for streamer and saltwater flies.

Sedge flies Flies tied to represent the Trichoptera, or caddis flies.

Setae Tails or whisks.

Slip A section of wing quill feather used for wings or tails.

Spent wing A wing style; the wings project horizontally, not vertically. Used on 'spinner' patterns.

Spey hackle Ultra-large feathers found on the sides of a cockerel's tail. Used on Scottish Spey wing salmon flies; similar to the Dee Wing flies. Both types are referred to as strip wing flies.

Spinner The final stage in the life cycle of the Ephemeroptera (the imago). Dead spinners are known by anglers as spent spinners or spent gnats.

Streamer flies Lures made with feathered wings often trailing beyond the bend of the hook.

Tandem fly A fly tied on two or more hooks linked in tandem.

Tippet Feather taken from the neck of a golden pheasant. Tippets from other pheasants are sometimes used.

Topping See crest.

Tube flies Salmon flies made from hollow lengths of plastic, aluminium, copper or brass tubing.
Underbody As the name suggests, this is a primary body applied before tying the final body of tinsel, etc. It builds up a definite shape.
Waddington shanks Straight hook-shanks used to form salmon flies. A treble is attached to the end. Tied and used in much the same way as tube flies.
Wet fly A fly tied so that it sinks below the water surface, imitating various aquatic creatures, or swamped terrestrials. Sometimes wet flies are attractors, imitating no specific creature but having plenty of flash and/or colour.
Wet fly hackle A soft-fibred hackle, usually from a hen's neck (not a cock's), used for wet flies.
Whip finish A method of finishing off the fly. It has a wrapping of at least three turns wrapped into itself.

COLOUR DEFINITIONS

Colour, like beauty, is in the eye of the beholder. One man's red is another man's brown. In some circumstances it is very difficult to describe colour in words. In fact, the languages of some cultures include very few colours, as those who have had flies tied in Africa know. Unless you send the actual fly, you are likely to get anything but what you asked for. Red, for instance, can mean anything from pale pink to deep purple.

The following list covers the hackle colours and natural shades a fly-dresser is likely to come across.

Black Good natural black capes are hard to come by. They can be anything from black to dark grey.
Chinchilla Pale barred Plymouth rock; grey on white.
Cree Barred feather: black, red and ginger; sometimes white flecked.
Ginger chinchilla Ginger bars on white.
Badger White or cream with a black centre.
Greenwell Ginger or light red with a black centre.
Coch-y-bonddu Medium to dark red with black centre and black tips.
Furnace Medium to dark red with black centre.
Light furnace As greenwell.
Grizzle Plymouth rock, barred black on white; once called cuckoo.
Dun Greyish brown, mousey.
Honey Pale creamy ginger.
Honey dun Light ginger with a dun centre.
Brassy dun Dun shade flecked with ginger.
Rusty dun Dun shade with red cast, similar to brassy dun.
Blue dun Slate coloured.
Iron blue Darker version of the blue dun, described as inky.
Brown Sometimes referred to as Coachman brown; brownish red.
Red Natural brown to almost ginger. These come from the Rhode Island Red breeds of fowl.
Dark red Sometimes called Old English Game.
Cream Off-white almost to honey.
White Good quality whites are rare. Second-grade and below are usually dyed to other colours.
Straw Cream with a definite ginger shade.
Olive dun A dun with a distinct olive hue. Very rare.
Badger dun Dun with a black centre. Very rare.

HOOK-SIZE CHARTS

The hooks depicted here are those that I use for my own fly dressing. The choice of hook is a matter of personal preference for most fly-tyers, but sometimes it must be tempered with availability. For this reason I have avoided being specific regarding type or make of hook in the main body of the book. I have given size only, except when the hook was of such a style as to be an important part of the overall shape of the fly; e.g., some nymph and shrimp patterns. Some, if not all, the hooks shown here should be available in any fly-dressing country.

Trout hooks (wet and dry)

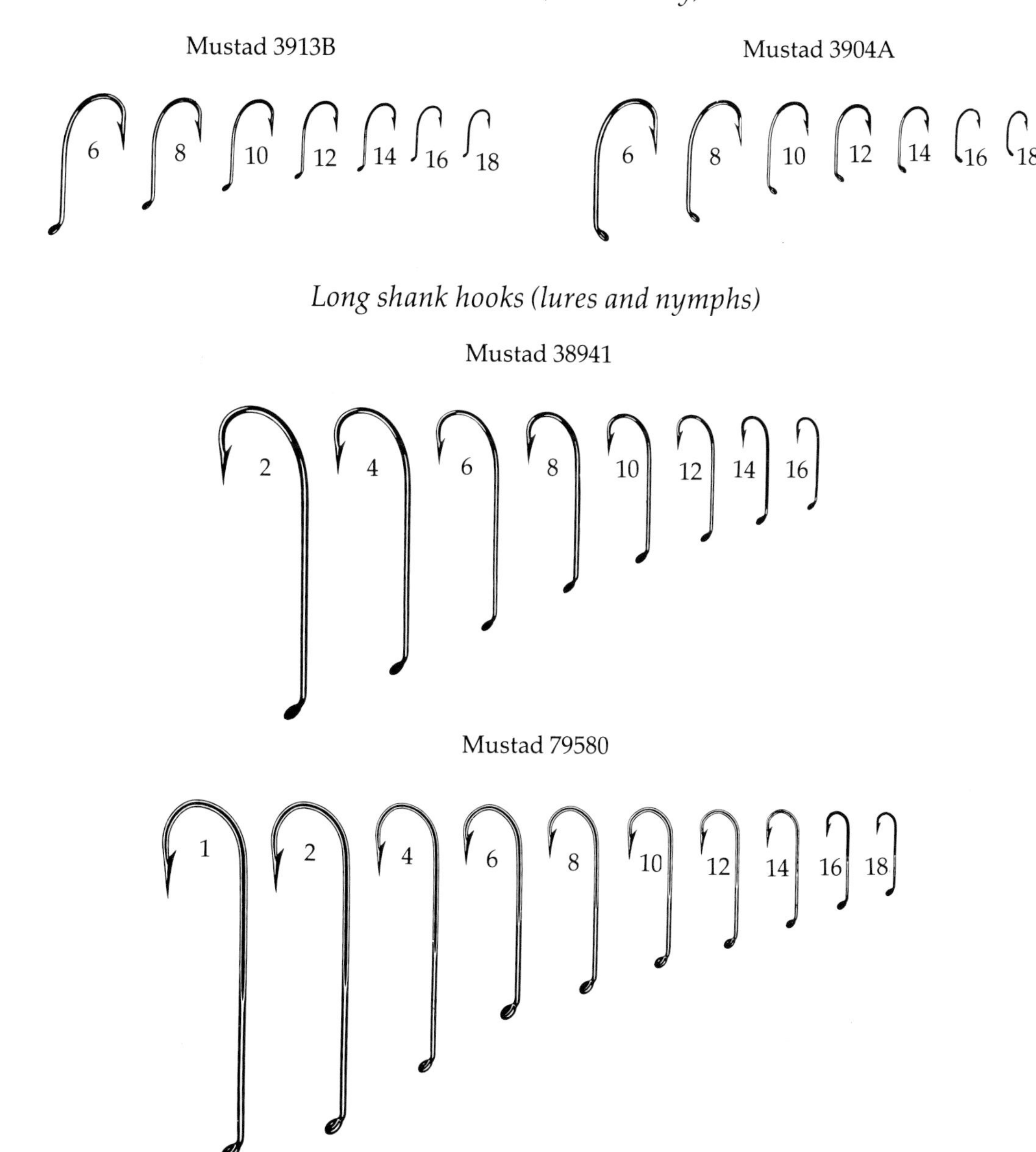

Special trout hooks

Partridge Shrimp/Grub Hook K4A

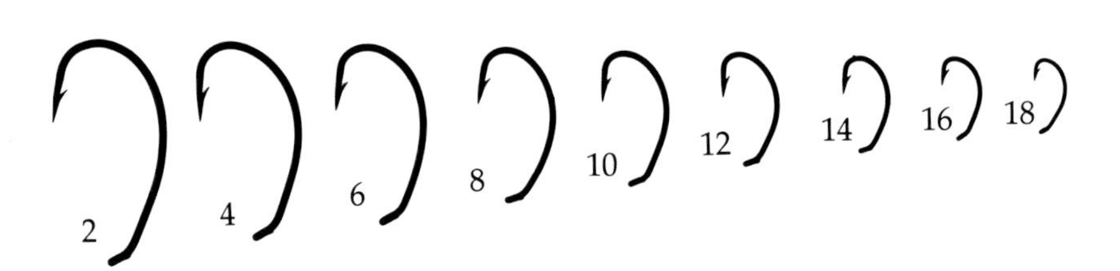

Partridge Yorkshire Sedge Hook K2B

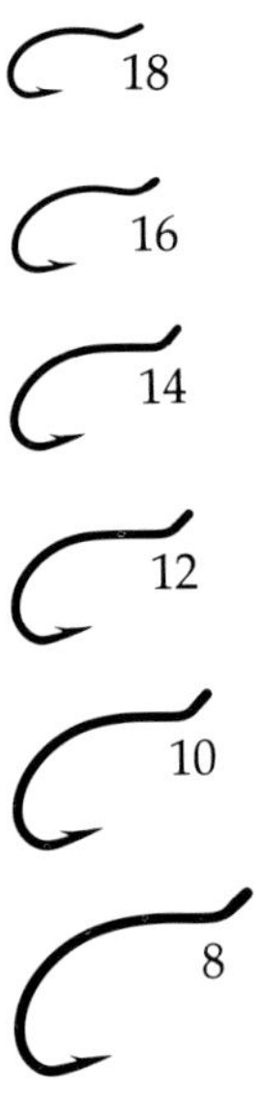

Mustad 37160 (Old English Bait Hook)

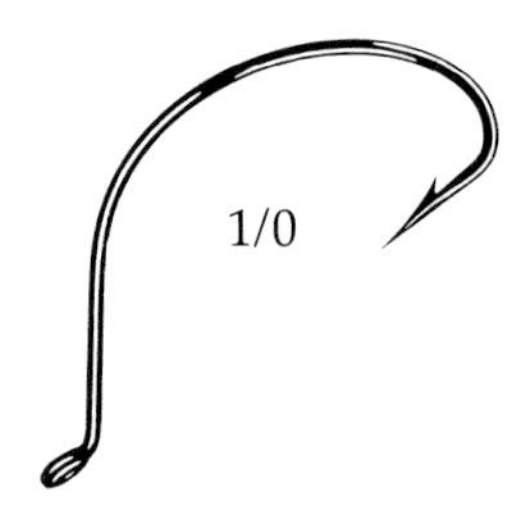

Partridge Barbless E3AY

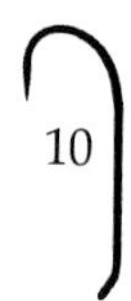

Mustad 3257B Barbless

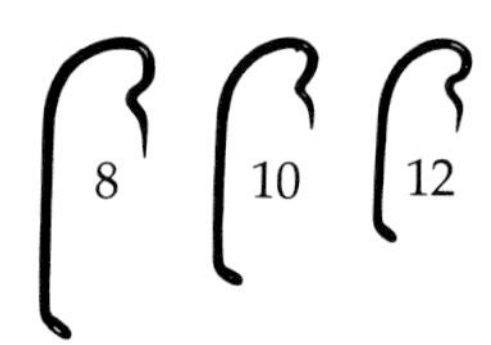

Salmon hooks

Mustad 36890 Single Salmon Dublin Limerick

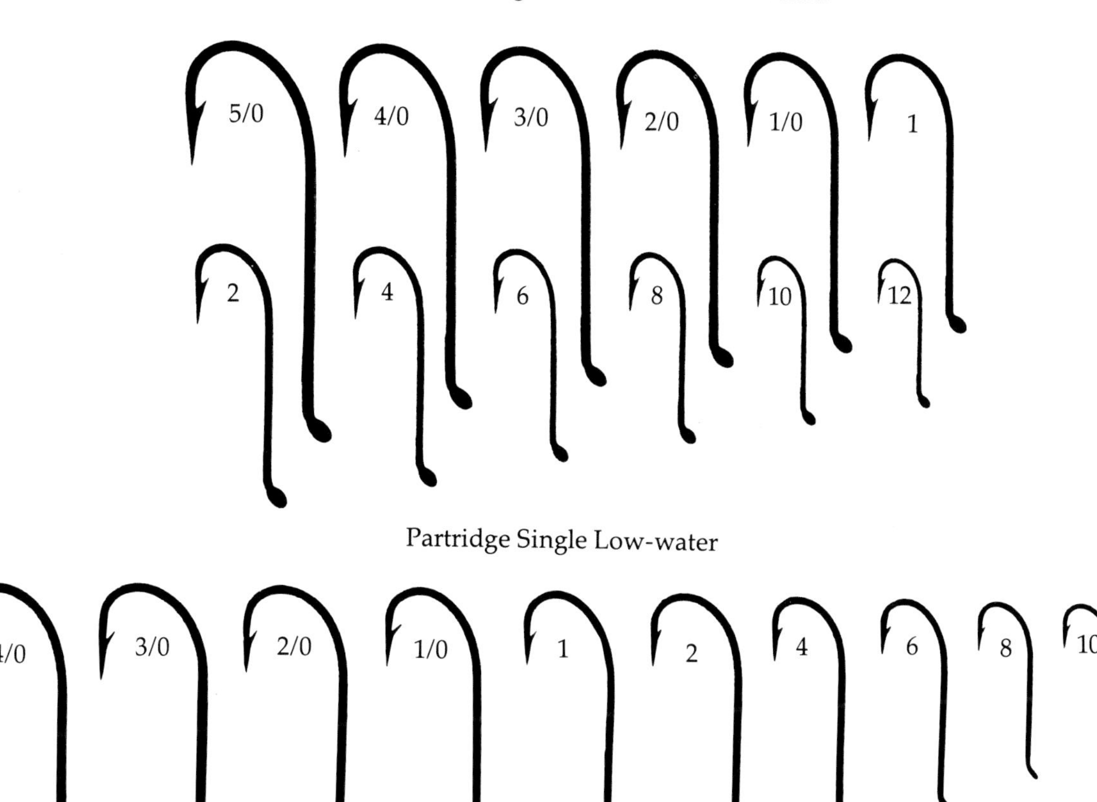

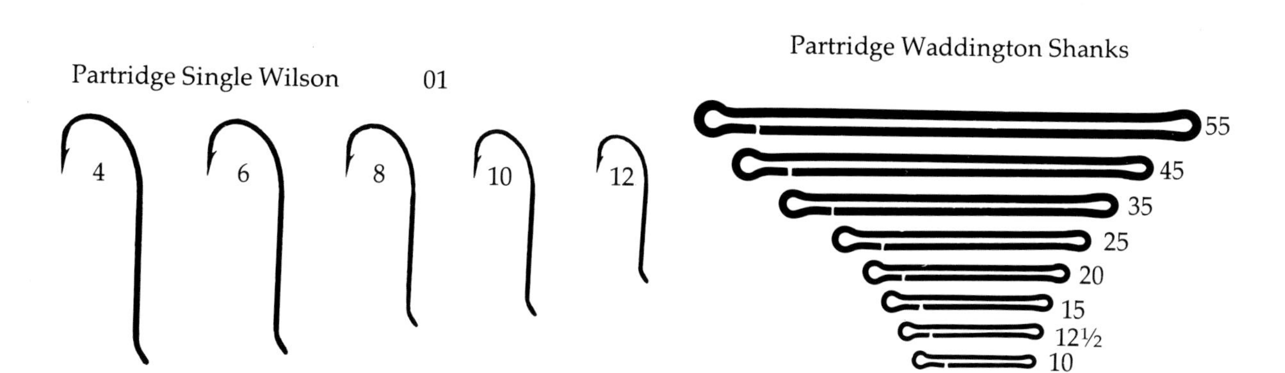

Partridge Double Low-water Code Q

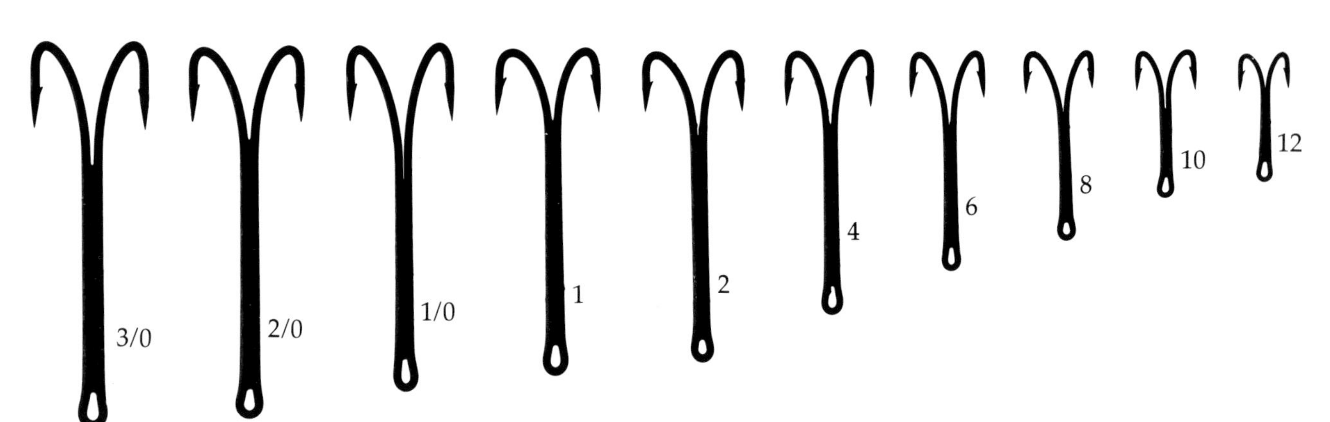

Partridge Double Salmon Code P

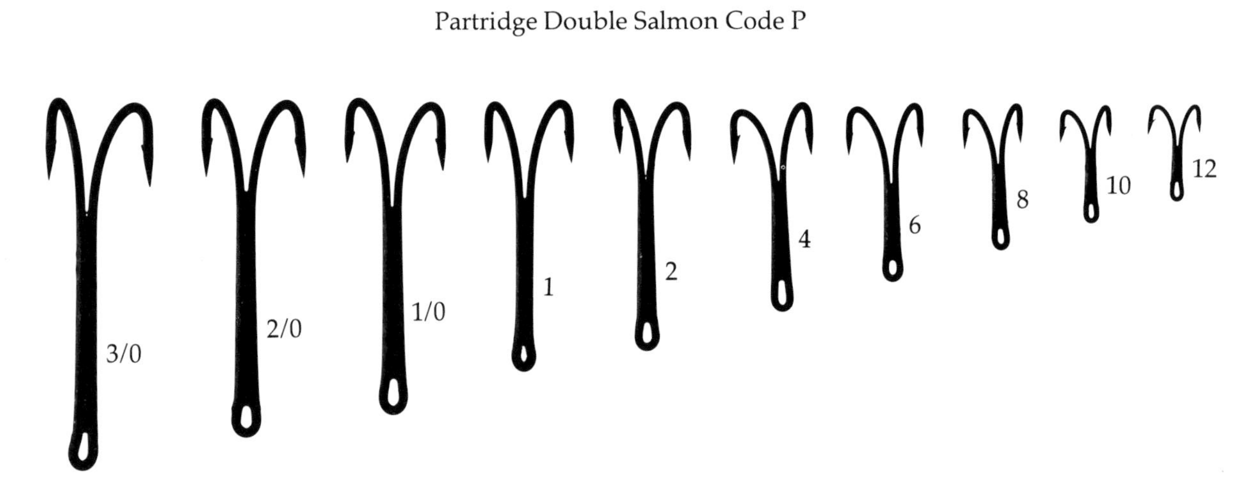

Partridge Long Shank Treble X2ST

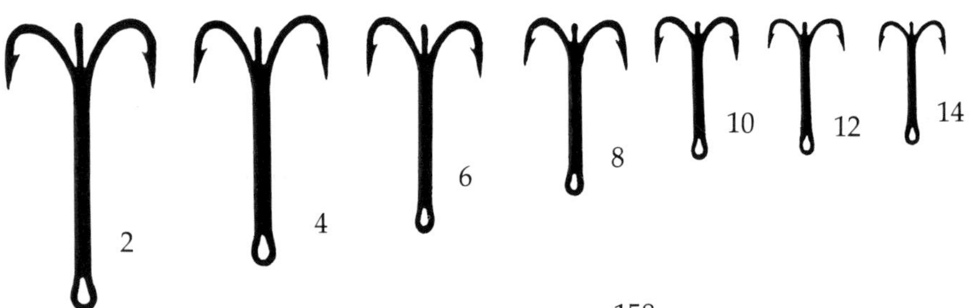

Index of patterns